Transnational Geographers in the United States

Transnational Geographers in the United States

Navigating Autobiogeographies in a Global Age

Edited by Alan P. Marcus

LEXINGTON BOOKS
Lanham • Boulder • New York • London

Published by Lexington Books
An imprint of The Rowman & Littlefield Publishing Group, Inc.
4501 Forbes Boulevard, Suite 200, Lanham, Maryland 20706
www.rowman.com

Unit A, Whitacre Mews, 26-34 Stannary Street, London SE11 4AB

Library of Congress Cataloging-in-Publication Data Available

ISBN 978-1-4985-0948-0 (cloth : alkaline paper)
ISBN 978-1-4985-0949-7 (electronic)

∞™ The paper used in this publication meets the minimum requirements of American National Standard for Information Sciences Permanence of Paper for Printed Library Materials, ANSI/NISO Z39.48-1992.

Printed in the United States of America

For the Amiable Baltimoreans

Contents

Preface

Alan P. Marcus

I came to know more about the connections between geography and autobiography after I took two graduate classes at the University of Massachusetts, Amherst, taught by Richard W. Wilkie ("Sense of Place," and "Spirit of Place"). Wilkie's approach encouraged his students to explore their respective personal geographical settings in an autobiographical narrative style. Later, I rewrote the final paper I had written in one of his classes and eventually published it as a book chapter. Today those frameworks I had learned from him bring me back here, to re-think *discoveries of place* through autobiography and geography, as they collectively project the narratives I have edited in this volume, or as I call them, *autobio-geographies*.

I want to thank the late Susan Hardwick, who helped bolster my research on transnational migration and, later, this project. I want to acknowledge her encouragement and my sadness to know about her passing in 2015. I had met Susan for the first time at a Race, Ethnicity and Place (REP) conference (Tom Boswell had introduced us), and over the years we became friends and shared research ideas. She had encouraged me enthusiastically to pursue this project idea when I first told her about it. She had kindly given me three books she owned—she thought they would be helpful for this particular volume (which they have been). And just recently in March, 2015, at the Association of American Geographers annual meeting in Chicago, we met and talked over lunch. I was shocked and saddened when I found out that she had passed shortly afterwards, unexpectedly in late 2015. We mourn her loss. Susan will be missed as a colleague, scholar, and a dear friend.

I am thankful to the assistant editor, Brighid Stone, and staff at Lexington Books for their help and assistance in publishing this volume. I want to thank my colleagues and friends from the Ethnic Geography Specialty Group (EGSG), Race, Ethnicity and Place conference organizers (REP), and the Association of American Geographers (AAG), who have supported and acknowledged my work since I was a graduate student, and who continue to support so many other geographers. It was through the EGSG and REP conferences that I came to know the authors of this book. I want to thank them for agreeing to write their respective chapters and participating graciously in this compelling project.

The bustling literary and poetry scene in Baltimore is stimulating. In this case, I want to thank Shirley Brewer, Michelle Kreiner, Anne Dugan, Barbara Morrison, Judith Krummeck, Rafael Alvarez, and Clarinda Harriss, who have inspired me with the writing process.

I am especially grateful to the College of Liberal Arts at Towson University for their support during my sabbatical Spring 2016. I am grateful to my colleagues and friends at the Department of Geography and Environmental Planning at Towson University, and wish to thank: Charles Schmitz, Paporn Thebpanya (Bee), Virginia Thompson, Kent Barnes, Jim Smith, Jeremy Tasch, Jay Morgan, Marty Roberge, Shou Lu, Sya Kedzior, Todd Moore, Natalia Fath, Wayne McKim, Jim and Kay Dilisio, and Terry Cooney. I also need to thank Josette and Derrick Marcus, Barbara and Viggo Anderson, Debbie Martin, Patricia Martin, Tina Martin, Richard Wilkie, George Roberson, Mary Wilson, Joya Misra, Maxine Margolis, David and Angela McCreery, Susan Hardwick, Tom Boswell, Charlie and Bernice Courtney, Eric Rose, Larry Fogelson, Herb Hollidge, John Strait, and David Strait for their support and academic collaboration. Without their warm friendship and encouragement, the completion of this manuscript would not have been possible.

Introduction

Alan P. Marcus

The "Cocktail Party Syndrome," facetiously dubbed by Peter Gould, is when somebody at a cocktail party approaches a geographer and asks: "So, what do *you* do?" (Gould 1985). As geographers, we tend to prepare ourselves in anticipation for that moment, as the general public is still baffled by *what geographers do*. A simple reply to such a question might likely raise eyebrows and trigger other questions, such as: "Is geography about memorizing the capitals of the world, rivers, mountains, and volcanoes?" or, "Oh, is that like geology?" The geographer seems to need to produce stock explanations readily available in order to calm the concerns of those who ask similar inquiries. Likewise, the transnational individual also faces similar situations and questions. Explaining their background or their elastic national, ethnic, religious, or cultural identity unfolds into complex stories, and often a simple reply will not fit into clear-cut frameworks typically conveyed in academic, political, or public discourses. In this case, transnationals and geographers *transgress* and *transcend* clear, defined boundaries. Given these structural flexibilities, here I want to explore how transnational geographers navigate their own personal and professional livelihoods. This volume is a collection of essays geared toward this endeavor.

Both the discipline of geography and the concept of transnationalism share innate flexibility, hybridity, and inter-relatedness. As Aihwa Ong aptly states: "*Trans* denotes moving through space or across lines, as well as changing the nature of something . . . transnationality also alludes to . . . the transgressive aspects of contemporary behavior and imagination" (1999, 4). Adding to Ong's notion of *trans*, here I question and challenge notions of *self* in terms of transgressing and transcending boundaries and borders—and I investigate the meaning of "home," sense of place, and/or sense of *displacement*.

In order to better understand how transnational geographers navigate their livelihoods, I introduce a semi-autobiographical framework to answer questions such as: "How did you end up here?"; "What is your geographical story?"; and "Where is *home*?"—I call this "autobiogeography" (autobiography + geography). I combine "geography" with "autobiography." Hence, *autobiogeography* here becomes the collective "stories of place(s)" as they translate the mapping of our trajectories geared to answer those questions above through narratives. Drawing from personal and professional experiences, here we better examine the concept of place(s) and transnationalism at the microlevel.

There are limitations and challenges undertaking such an endeavor. These essays do not fit within the conventional scholarly format (e.g., with a theory and methodology section); the unorthodoxy here is intentional. The current literature on transnationalism is quite extensive and diverse (e.g., see Khagram and Levitt 2008; Friedman and Schultermandl 2011), and it is not my intention to systematically examine the multifaceted theoretical dimensions of transnationalism; therefore, this book does not cover the wide breadth of transnational literature.

The authors in this volume were faced with the challenging decision about what information to include or exclude in the recounting of their personal or professional lives. Yet at the same time, they still maintain rigor and validity by using the "I" voice to enhance the narrative about place through a personal angle. However, the intention is not to project these narratives as a formal autobiography or a memoir. I have not included maps or figures here due to budget and space constraints. Nevertheless, this undertaking is still a geographical project. Perhaps this volume will dovetail as a predecessor with a future publication for other autobiogeographies that will include maps, visuals, and photos—long overdue in a forthcoming project. I believe that a project with those dimensions would enhance the concept of autobiogeography, and would provide potentially creative pedagogical as well as academic outlets.

As Pamela Moss reminded us: "Autobiography is one of the most fruitful ways to access the process of building a discipline" (2001, 7). In this case, the intention here is to highlight the autobiogeographies of eight selected geographers who are university faculty members, and who work and reside in the United States. They have conducted and published their research work on immigration and transnationalism, and are also at the same time, transnational migrants. I wanted to know more about their own "stories." The authors conspicuously and unapologetically write here with a capital "I" voice. I encouraged the contributing authors to use "I" as much as they wished, in order to capture, enhance, and highlight the personalized account of their experiences. Hence, some of the essays take a wide range of stylistic directions.

The authors in this volume have spent most of their careers researching immigration, however, until now there has been little or no attention given to better understand their own migration story or how they "ended up here"; and, especially, how they have experienced geography and transnationalism. This volume will be of interest to other geographers as well as to other scholars of social sciences and humanities and the general public, because of their devotion to the concept of contemporary transnationalism, borders, fragmentation, movement, location, space, and especially, the qualitative concept of place.

It is well known that the social science literature covering U.S. immigration and transnationalism has typically focused on lower socioeconomic populations. It is well established that the public treatment of the term "immigrant" has been and still is inflammatory in political rhetoric in the past and present. However, very little or no research utilizes the narratives of university faculty—or in this case, geographers—who are themselves U.S. immigrants (with the exception of the volumes cited below). As far as I know, only four books are currently available that address similar topics and questions on transnational geographers. However, none of them address the trajectories of transnational geographers *in* the United States.

For example, Remy Tremblay and Susan Hardwick (2014) edited an important volume, "Transnational Borders and Transnational Lives: Academic Mobility at the Borderland," which addressed the biographies of U.S.-born geographers in Canada and the concept of transnationalism. Similarly, Leslie J. King (2007) edited "North American Explorations: The Memoirs of Geographers from Down Under," which looked at the trajectories of Australian geographers in the United States, written in the style of memoir. Pamela Moss edited "Placing Autobiography in Geography" (2001), a collection of autobiographies written by international geographers, including Anne Buttimer and Pamela Moss. And finally, Peter Gould and Forrest R. Pitts edited "Geographical Voices: Fourteen Autobiographical Essays" (2002), also a collection of autobiographies of world-renowned geographers (all men), such as Yi Fu Tuan, Gilbert White, David Harvey, Richard Morrill, and Waldo Tobler. All of these essays were written in the form of autobiography or memoir. The perspectives of women in geography are not included—in fact, not a single woman—a fact that Gould and Pitts (2002) pointed out in the preface. None of these books, however, exclusively address the trajectories of transnational geographers who reside and work in the United States. This volume will fill this gap, as transnational geographers become more common fixtures in U.S. geography departments, and as geography becomes increasingly a vital topic of interest to the general public and U.S. students.

As the editor of this book, I also have strived to show how to accommodate changing disciplinary and national perspectives on immigrants and geography, and intend to show how these perspectives portray important adjustments within contemporary global demographic trends, which in turn, reflect ever-changing identities. Therefore (by proxy), these perspectives also help to promote U.S. global competence.

How do eight geographers who live in the United States personally and professionally navigate their lives? How do transnational geographers understand *self* in relation to the public, or the academy? Where is "home"? Using autobiogeographical approaches, the authors here will answer these questions. I invited eight professional geographers who are also active members of the Association of American Geographers (AAG), Ethnic Geography Specialty Group. This is the first publication about their own trajectory—their autobiogeography. They share this navigation process through narratives that each author has with a respective place. In chapter 1, Alan P. Marcus (Brazil); chapter 2, Wei Li and Shao Lu (China); chapter 3, Ibipo Johnston-Anumonwo (Nigeria); chapter 4, Heike Alberts (Germany); chapter 5, Heather Smith (Canada); chapter 6, Stavros Constantinou (Cyprus); and chapter 7, Weronika Kucek (Poland). As Margaret Walton-Roberts stated, "Geographers are eminently positioned to make a contribution to this debate, but have been slow to develop an explicitly transnational focus" (2003, 237). Therefore, this volume taps into a new and different understanding by using autobiogeography to better understand transnationalism "inwardly": personally and professionally. Instead of explaining and summarizing each chapter, I would rather have the reader interpret and *navigate* the chapters on their own terms. Since each chapter differs widely in style as well as in terms of geographical location, some have been more reflexive, others more systematic in terms of time and duration, while at the same time the goal all along is to provide enough flexibility to encourage different styles of discussing and interpreting autobiogeographies.

The transnational individual is born in one place, lives in another, yet lives in both at the same time, and continually maintains multiple loyalties, traversing two or more places concurrently over time. Again, Ong states: "The multiple-passport holder is an apt contemporary figure; he or she embodies the split between state-imposed identity and personal identity caused by political upheavals, migration, and changing global markets" (1999, 2). Like the geographer who transcends and transgresses strict disciplinary boundaries, the transnational individual *transcends* national boundaries and, hence, *transgresses* the constraints of living their lives within explicit national borders. This edited volume of collected essays illustrates the trajectories of selected transnational geographers who navigate their various complex identities in a global age.

WHY IS THIS PROJECT IMPORTANT? WHAT IS GAINED FROM THIS KNOWLEDGE?

In an age of political soundbites, where U.S. public discourse on racial identity is typically framed within binary racial and political platforms (i.e., black/white optic); therefore, a giant void appears in public discussions about other dimensions of identity and of complexity. Only recently, the 2010 U.S. Census offered a new category of "multi-racial." These discussions are not about race. However, I want to begin this section with the premise that most of the world is (nor has been) neither black nor white in the clear-cut manner their discussants intuitively suggest. Most of the world is typically far more complex than the U.S. racial binary optic usually used to frame discussions on identities and categories (e.g., the populations of India and China have surpassed the one billion mark, and Latin American populations do not use clear-cut white-black categories and so on). Therefore, the transnational individual transcends notions of category boundaries that are typically used in U.S. racialized discussions.

Research within the social sciences and cultural studies on immigration and ethnicity make heavy use of metaphors (e.g., "melting pot," "ethnic quilt," "ethnic mosaic," etc.). However, we should be cautioned, as these metaphors may mislead more than they inform (e.g., Marcus 2009a, 2009b). Given that each individual immigrant experience differs widely on many levels (e.g., spatial, national, temporal, linguistic) the collective understanding of immigrant groups neglects these many differences within migration processes at the macro and micro levels—and this is particularly the case with transnational migrants. Here I explore how and why the trajectories of transnational geographers can be valuable to further examining the concepts of borders, boundaries, and nationhood.

First, the transnational geographer is adequately well-equipped to inform us firsthand about the politics of identity and foreign languages. Moreover, the essays in this volume allow us to interrogate concepts and geographical ascriptions that challenge ethnic, national, and cultural labels at the micro level. Therefore, the authors in this volume share multiple interconnected strands from various complex national, regional, ethnic, religious, and cultural backgrounds ranging from Brazil, China, Poland, and Canada to Nigeria, Germany, and Greece. Second, this volume follows the effort to accommodate a global society which is becoming increasingly hybrid, epitomized by President Barack Obama, for example, and by the 2010 U.S. Census newly created category for "multi-racial"—as mentioned earlier—conforming to the trend to acknowledge multicultural and multiracial realities in the United States.

Third, this volume includes the trajectories of women in geography—a discipline where women are still disproportionately under-represented. This volume helps to fill that gap (all but two chapters were written by women). The intention is to also acknowledge and reflect the importance of women within geography and their contemporary contributions to the discipline.

Finally, the stories of transnational geographers illustrate how disciplinary perspectives are increasingly being produced by authors whose personal and professional backgrounds are more complex than their own names might indicate at first glance. Furthermore, one could argue that the very etymology of geography ("writing of the Earth") lends itself to a literary, descriptive, and qualitative slant. The narratives in this volume provide us with an opportunity to highlight the personal and professional complexities of transnational geographers.

Whither Geography?

Richard Morrill had discussed the importance of migration within geography: "Migration is indeed a quintessentially spatial social process, in which the environmental, place, and space components of geography are essential to a full understanding" (1987, 536–537). By proxy, transnationalism is quintessentially a spatial social process, and thus it is *about geography*. The concept of transnationalism maintains multiple economic, sociocultural, and national ties and loyalties to two or more countries simultaneously (e.g., Basch et al. 1995; Levitt 2004, 2007; Marcus 2009a, 2009b). Furthermore, Peggy Levitt aptly explains: "The assumption that people live their lives in one place, according to one set of national and cultural norms, in countries with impermeable national borders, no longer holds" (Levitt 2004, 1). This quote is perfectly situated here. In recent cultural studies and social science research, the use of "transnationalism" has drawn upon previous economic meanings on globalization processes, and its focus today tends to refer to the diminished significance of national borders in the production and diffusion of ideas, objects, and people (Basch et al. 1995, 48).

Social activities and consumer behavior ranging, for example, from retrieving information from the World Wide Web and sending electronic mail to drinking Coca-Cola, wearing jeans, listening to pop music, and eating at McDonald's are no longer restricted to one single geographic location or culture, and as Peggy Levitt informs us, "the spread of global media has brought the 'core' to all the world's 'peripheries.' Many of today's migrants arrive partially socialized into aspects of Western, if not North American culture" (2001, 25). Hence, many of today's migrants are already *Americanized* before they arrive in the United States. Transnationalism offers an alternative approach to linear models that have been used to examine past human migration patterns. Instead of looking at merely permanent or temporary migration processes, the strength of transnational approaches is that it provides additional and alternative views on migration dimensions.

Transnational flows of information occur beyond the boundaries of a formalized territory, hence, they are *meta-territorial*. These meta-territorial interactions and relations occur in ways that are different from previous sociocultural practices and processes enacted by immigrants in the United States decades ago. That is, today sociocultural shedding of loyalties and ties to place of origin, as well as sociocultural references of origin, do not occur in the same ways (intensity, scope, and scale) as previous immigrant groups who arrived one hundred years ago, precisely because of the advent, access, and velocity of recent technological advancements in telecommunications (e.g., cellular phones, online world news, weblogs, videos, and/or other modes of telecommunications in general) as well as access to and velocity of modes of transportation. Current research shows how and why recent transnational migration considerations are different from the frameworks that had been used in the past to understand migration movements (i.e., Levitt 2004; Marcus 2009a, 2009b, 2010, 2011b, 2011c, 2011d, 2013b; Margolis 1994, 1995, 2013; Schiller et al. 1995; Tsuda 1999; Walton-Roberts 2003).

Saulo Cwerner has also suggested that migrations carry crucial temporal dimensions, but times also *migrate* with people: "The time perspectives and symbols of immigrants affect in many ways their predicament in the 'host' society" (2001, 7). The ease of communication networks and access allows newly arrived immigrants to pursue contacts for potential employment, social networks, and/or housing once they have arrived in the place of destination. The velocity and availability to access information and communication flows facilitate the increased geographical mobility of migrant transnational processes.

Within the social sciences, the use of the term "transnationalism" draws upon previous economic meanings based on globalization processes, and the term usually refers to the diminished significance of national borders in the production and diffusion of ideas, objects, and people (Schiller et al. 1995, 48). Transnationalism offers an alternative approach to linear models that have been used to examine past human migration patterns. As Peggy Levitt and Mary Waters describe, transnationalism is "how ordinary individuals live their everyday lives across borders and the consequences of their activities for sending-and receiving-country life" (2002, p. 8). Instead of looking at merely "permanent" or "temporary" migrants, the strength of transnational approaches provides an alternative view on migration dimensions. Transnational approaches look at processes by which migrants "forge and sustain multi-stranded social, economic, and political relations that link together their societies of origin and settlement" (Basch et al. 1994, 6).

Maxine L. Margolis fittingly pointed to the importance of multi-dimensional approaches to understand Brazilian transnational migration processes (2013, 1998, 1995, 1994), particularly in her groundbreaking research (1994). She explained: "The traditional 'push-pull' explanations of international migration have proved inadequate in accounting for a worldwide phenomenon of such magnitude . . . they tend to ignore macrostructural factors that enmesh these global movements" (1994, xv). Margolis pointed to other push/pull factors, such as "rising expectations in sending countries," and how the globalized world of technology emerges as a significant disseminator of the "information factor." Furthermore, Castles and Miller explain: "Macro-, meso- and micro-structures are intertwined in the migratory process, and there are no clear dividing lines between them. No single cause is ever sufficient to explain why people decide to leave one country and settle in another" (2003, 28). Transnational migration processes and practices affect multiple social, political, and economic activities where ties, connections, and loyalties engaged between country of origin and country of destination occur beyond formal borders. For example, in his study on Brazilian immigrants in Japan, Tsuda states: "Many migrants are neither temporary nor permanent, but remain in a *liminal* state between the two extremes, uncertain about their futures and their country of residence" (1999, 692).

Transnational flows of information occur, then, beyond formalized territory and boundaries; hence, "International migration can no longer be seen as a one-way process. Events, communities, and lives, most observers now recognize are increasingly linked across borders" (Levitt and Waters 2002, 5). These meta-territorial interactions and relations are unique today with the advent of recent technological advancements in telecommunications, the Internet, and jet travel.

The Condition of Being a Geographer

In most U.S.-based geography (or geosciences) departments, it is well known that the range of research interests is so broad and widely defined that sometimes the schism within research interests can make them appear so far apart from each other, they might seem they are produced from different departments altogether. However, all geography students in training benefit greatly from taking liberal arts courses (and I would add that they *should* take a wide-ranging variety of liberal arts courses), as Carl Sauer had once recommended (Mathewson and Denevan 2009). Most geographers today agree that both quantitative and qualitative approaches are just as valid and rigorous, if designed and completed effectively: "Assertions made in the 1970s that qualitative research was 'unscientific' were no more credible or creditworthy than assertions made in the 1980s and into the 1990s that quantitative research had nothing to offer" (Hoggart et al. 2002, 306). Using unidirectional approaches may create the illusion that science is produced in a cultural vacuum and is, by proxy, devoid of human agency. So is the geographer an artist or a scientist, or both?

The various and wide-ranging methodologies within the disciplinary practices of geography create a sense of *disciplinary transgression*—yet precisely because of this transgression, and ironically at the same time, the geographer is aptly suited as an innovative researcher. For example, all research, regardless of what kind, broadly speaking but not necessarily in this order, goes through the following six phases of research: (1) start with an issue or problem, (2) examine the literature in some way related to the problem, (3) pose questions, (4) gather data, (5) analyze them, and (6) write up the reports (Creswell 2007, 41). Thus, it is well known among us geographers that the question the researcher must ask is, "Which methodology best fits the research question?" and not, "Which methodology is more scientifically more valid, or better?" In other words, the researcher needs to first find which methodology will be more effective in answering the research question at hand. Henceforth, it is easy to see how geographical research spans widely in terms of approaches, and may disrupt and transgress clear-cut disciplinary boundaries.

David N. Livingstone (2003) showed how science and its perceived objective and omnipotent endeavor are, in fact, much to the contrary; *subjectively* predicated upon spatial and temporal circumstances. Our various knowledge claims are indeed contingent on place and time, and become salient in research writing, as Berg and Mansvelt state:

> The process of writing constructs what we know about our research but it also speaks powerfully about who we are and where we speak from . . . the detached third-person writing style so common in academic journals and reports implies that the researcher is omnipotent. . . . Knowledge does not . . . exist independently of the people who created it—knowledges are partial and geographically and temporally located. (Berg and Mansvelt 2000, 173)

In this case, the geographer is an unconventional scholar on two levels: (1) *Artistic:* That is, he/she transgresses and transcends the monolithic approach to understanding the *aesthetics of place* through various approaches toward place-description, including the usage of humanities-driven approaches as well as maps, drawings, photography, graphic, visual, and spatial thinking. Maps have been compared to the art of painting, in terms of their aesthetic dimensions. The artistry does not stop with maps; as the art of the geographer also involves the description of the *aesthetics of place* (e.g., Tuan 1974, 1976), also seen in the geographical tradition. Maps are also aesthetic tools for political means (e.g., Livingstone 1994, 2003)—they are political tools operating on perceptual as well as conceptual levels. Therefore, geographers reach artistic spheres that lay beyond the visual realm of mapmaking; the artistry is also attained through literary sensitivity, through the *poetics of place,* and through the descriptive endeavor of informing the significance of place and experience in a multitude of literary dimensions.

(2) *Scientific* in terms of methodology, theory, and fieldwork research (i.e., geography does not have a unified disciplinary methodology like physics or chemistry, for example). The language geographers use—their tools of the trade—can vary to such a degree, that a non-geographer could easily read a professional geography journal but think that she/he just read several articles from different disciplines in one volume. The discussion on whether to stress the scientific method within geography has been extensively discussed already. In the 1950s/1960s, for example, discussions about the future of geography (and the debate of what it *is*)

included whether the discipline should be exclusively a quantitative-based science—an idea which no longer holds.

The discipline of geography has gone through periods of self-doubt in relation to other disciplines in academia. In the past 60 years or so, geographers may be self-described (or perceived by others) as scientific *unconventionals* because geographers have traditionally seen the world through various theoretical/methodological lenses of analysis. The elimination of geography at Harvard University in the late 1940s (in which, later, other top-tiered U.S. universities followed suit and also eliminated their geography departments) perhaps exacerbated this academic self-doubt, and pushed geographers to further seek a unifying method and theory, which in turn caused further internal tension and anxiety within the discipline. However, this self-doubt was misguided.

These two levels of articulating the concept of geography as *unconventional* (i.e., transgressing and transcending) are inter-related. In this case, the discipline of geography can be understood in contradictory terms: as an artistic and as a scientific enterprise, at the same time—this would make sense, if we take the premise that modern European and U.S. geography inherited the legacy from ancient Greek literary and mathematical traditions (i.e., in modern vernacular this would be loosely equivalent to the *ideographic*/descriptive/qualitative approach combined with the *nomothetic*/"law-seeking"/quantitative approach, respectively). But how can a modern discipline (especially if viewed as a "scientific" one) within the academy not have a unified theory or method? And how can it be both science and art? According to the unidirectional and traditional scientist, this cannot be. However, as Creswell aptly states: "Qualitative inquiry represents a legitimate mode of social and human science exploration, without apology or comparisons to quantitative research" (2007, 11). While most other academic disciplines typically retain rigid theoretical and methodological traditions, with clear-cut disciplinary boundaries, geography is porous and multi-disciplinary in its theoretical and methodological approaches, and it should cross-pollinate with other disciplinary approaches.

The objective here is to articulate the fragmented narratives of eight authors through geography and transnationalism on two strategic levels: (1) personal and (2) professional. The geographer works within a three-pronged disciplinary approach: social sciences, humanities, and physical sciences—situated sometimes as a science, sometimes as an art, or somewhere inbetween the two, or both at the same time, as science and art. Here eight transnational geographers discuss their own trajectories and perspectives on geography, transnationalism, and their autobiogeographies, as they each answer the questions such as: "Where is home?" and "How did you end up here?"

REFERENCES

Basch, L., N. G. Schiller, and C. S. Blanc. 1995. From Immigrant to Transmigrant: Theorizing Transnational Migration. *Anthropological Quartely* (1): 48–63.

Berg, L. and J. Mansvelt. 2000. Writing In, Speaking Out: Communicating Qualitative Research Findings. In *Qualitative Research Methods in Human Geography,* Iain Hay (Ed.), 161–182. Oxford, U.K.: Oxford University Press.

Buttimer, Anne. 1976. "Grasping the Dynamism of Lifeworld" Annals of the Association of American Geographers, no. 66: 277–292.

Denevan, W. M. and K. Mathewson. 2009. *Carl Sauer: On Culture and Landscape. Readings and Commentaries*. Baton Rouge, LA: Louisiana State University Press.

Friedman, M. and S. Schultermandl (eds.). 2011. *Growing Up Transnational: Identity and Kinship in a Global Era*. Toronto, Canada: University of Toronto Press.

Gould, Peter. 1985. *The Geographer at Work*. Routledge and Kegan Paul: NY.

Gould P. and F. R. Pitts (eds.). 2002. *Geographical Voices. Fourteen Autobiographical Essays*. Syracuse University Press: New York.

Hay, I. (ed.). 2000. *Qualitative Research Methods in Human Geography*. Oxford, United Kingdom: Oxford University Press.

Hoggart, K., L. Lees, and A. Davies. 2002. *Researching Human Geography*. London, United Kingdom: Arnold Publishers.

Khagram, Sanjeev and Peggy Levitt (eds.). 2008. The Transnational Studies Reader. Routledge, New York.

King, Leslie J. 2007. *North American Explorations: The Memoirs of Geographers from Down Under*. Trafford Publishing, Canada.

Levitt, P. 2001. *The Transnational Villagers*. Los Angeles and Berkeley: University of California Press.

———. 2004. Transnational Migrants: When "Home" Means More Than One Country, *Migration Information Source* http://www.migrationinformation.org/Feature/print.cfm?ID=261.

———. 2007. *God Needs No Passport. Immigrants and the Changing American Religious Landscape*. New York: The New Press.

Levitt, P. and M. C. Waters (eds.). 2002. *The Changing Face of Home: The Second Generation*. NY: Russell Sage Foundation Press.

Livingstone, D. N. 1992. *The Geographical Tradition: Episodes in the History of a Contested Enterprise*. Malden, MA: Blackwell Press.

———. 2003. *Putting Science in Its Place. Geographies of Scientific Knowledge*. Chicago, IL: University of Chicago Press.

Ong, Aihwa. 1999. Flexible Citizenship. The Cultural Logics of Transnationality. Duke University Press: Durham and London.

Marcus, Alan P. 2015. Convenient Liaisons: Brazilian Immigration/Emigration and the Spatial-relationships of Religious Networks, *Space Populations Societies* (2014-2-3) 2–1.

———. 2013a. Sex, Color, and Geography: Racialized Relations in Brazil and its Predicaments. *Annals of the Association of American Geographers* 103(5):1282–1299.

———. 2013b. Global moves from and to Brazil. In *Careers without Borders: Critical Perspectives,* 93–106, edited by Y. Baruch and C. Reis, Oxford, UK: Routledge.

———. 2011a.Transnational Rio de Janeiro: (Re)Visiting Geographical Experiences. In *Growing Up Transnational: Identity and Kinship in a Global Era,* 21–35, edited by May Friedman and Silvia Schultermandl. Toronto, Canada: University of Toronto Press.

———. 2011b. *Towards Rethinking Brazil: A Thematic and Regional Approach*. NJ: John Wiley & Sons, Inc.

———. 2011c. Rethinking Brazil's Place within Latin Americanist Geography. *Journal of Latin American Geography* 10(1): 129–147.

———. 2011d. Experiencing Ethnic Economies. *Journal of Immigrant and Refugee Studies* 9: 1–24.

———. 2010. Back to Goiás and Minas Gerais: Returnees, Geographical Imaginations and its Discontents. *Revista Tempo e Argumento* 2(2): 121–134.

———. 2009a. Brazilian Immigration to the United States and the Geographical Imagination. *Geographical Review* 99 (4): 481–498.

———. 2009b. (Re)Creating Places and Spaces in Two Countries: Brazilian Transnational Migration Processes. *Journal of Cultural Geography* (26)2: 173–198.

Margolis, M. L., 2013. *Goodbye, Brazil: Emigres from the Land of Soccer and Samba.* By Maxine L. Margolis. Madison: Wisconsin University Press.

———. 1998. *An Invisible Minority: Brazilians in New York City.* New York: Simon and Schuster.

———. 1995. Brazilians and the 1990 United States Census: Immigrants, Ethnicity, and The Undercount. *Human Organization* 54: 52–59.

———.1994. *Little Brazil: An Ethnography of Brazilian Immigrants in New York City.* Princeton, New Jersey: Princeton University.

Moss, Pamela (ed.). 2001. *Placing Autobiography in Geography.* Syracuse University Press: New York.

Tremblay, R. and S. Hardwick (eds.). 2014. *Transnational Borders and Transnational Lives: Academic Mobility at the Borderland.* Press de L'Universite de Quebec, Canada.

Tuan, Yi-Fu. 1974. Topophilia: A Study of Environmental Perception, Attitudes, and Values. New Jersey, NJ: Prentice-Hall.

———.1977. Space and Place: The Perspective of Experience. Minneapolis, Minnesota: University of Minnesota Press.

Tsuda, T. 1999. The Permanence of "Temporary" Migration: The "Structural Embednessness" of Japanese-Brazilian Immigrant Workers in Japan. *The Journal of Asian Studies* 58 (3): 687–722.

ONE

Navigating My Autobiogeography

From Brazil to Baltimore

Alan P. Marcus

During the summer of 2014, I was invited by Baltimorean writer and poet Barbara Morrison to be a guest speaker at Litmore (a literary center in Baltimore) together with local classical music radio WBJC broadcaster Judith Krummeck, who is South African–born. The two of us were invited to read excerpts from our respective publications. A few weeks before our talk, Judith had asked me: "How did you end up here?" I was taken aback by this simple question. I had found that stock answers such as "job, etc." usually made it easier for others to digest and understand. However, I also knew that stock answers were ultimately disingenuous. I started to look at the particulars in my life which led me to make the ultimate decision to leave Brazil, and proceeded to answer her question. I jotted them down and added vignettes, and eventually I attached them to a draft of this chapter. Like others who leave their home country to venture into another, we all leave some kind of "unfinished business" behind.

I will provide glimpses of my experiences in particular place(s) at different points in time, which I call autobiogeography. As Yi-Fu Tuan states, humanistic geographers strive to "understand how geographical activities and phenomena reveal the quality of human awareness" (1976, 627). And similarly, I explore my own fragmented geographical activities through my awareness of *self.* This chapter is not an autobiography or a memoir. I have explained my interpretation of and my intentions of using the term "autobiogeography" (i.e., autobiography + geography) in the introduction section of this volume, which will suffice, and I do not feel that I need to elaborate any further on its meaning.

In line with J. Nicholas Entrikin's statement "place, to the humanist, is not a collection of empirically observable objects and events, but rather is a repository of meaning" (1976, 626). In this case, I will examine the repository of place(s) and their meanings—my experiences from Brazil to Baltimore. Here, I have not addressed race, class, and gender or the socio-economic upheavals of Brazil. Yet it is not out of negligence or indifference, since I have explored and published those issues elsewhere (see Marcus 2009, 2011b, 2015). I have also discussed the theoretical backdrop of growing up transnational in a previous publication (Marcus 2011a), where I proposed a theoretical model to understand sense of place. Here I have added new dimensions within a new framework. I will explore facets from my personal experiences in the form of vignettes and personal snippets.

GEOGRAPHICAL IMAGINATIONS: LEAVING BRAZIL

I left São Paulo, Brazil, on a rainy November night in 1992. I remember my mother sitting next to me in the living room on that evening before I left for the airport. As I talked to her she looked at me teary-eyed. She intuitively knew that I would never be returning home. Once I boarded the airplane, I knew I had made a permanent life transition. I was heading to New York City to stay with a friend of mine, and I intended to carry on to Chicago to reconnect with bluesman Junior Wells (who had invited me to visit him as a guitarist). I believed I would eventually end up in the Marquesas Islands in the South Pacific, though I was not sure exactly how I would get there—but I was sure of one thing: I would leave Brazil and never live there again.

My mother owned a Paul Gauguin print copy, which hung on her bedroom wall. Those images of mountains and semi-dressed women fed my geographic imagination with utopian ideals I once held; hence the South Pacific seduced me at that time, and perhaps that is why I believed I would end up there eventually. I never dreamed that I would end up in Baltimore.

Despite my fondness for Rio de Janeiro—my place of birth—and despite the *sense of place* I experienced in Rio, I needed to leave Brazil. I grew up in Brazil during the military dictatorship (1964–1985), and, the 1980s was dubbed Brazil's so-called "lost decade," due to a national economic downturn and hyperinflation. I had little ambition to work in a country which suffered (and still does) from a chronic corrupt cultural malaise (as I write this chapter, Brazil is experiencing the same corruption problems and the worst recession in twenty years). I was not sure if there was something wrong with me or with Brazil, or with both.

Of course, since that period, Brazil has become a global participant (e.g., G-20 group, Mercosul, and BRICS: "Brazil, Russia, India, China and South Africa"), has hosted the World Soccer Cup in 2014, and will host the Rio 2016 Olympics; it has also become a regional area of interest to the world. Yet, as I had pointed out before (Marcus 2011a, 2013), within academic and public spheres surprisingly little attention has been given to non-Amazonian urban Brazil, with few exceptions. Given Brazil's enormity in terms of territory and population size (the largest in Latin America), this would be tantamount, for example, to ignoring China within Asian regional studies.

I never reached Chicago or the Marquesas Islands (I only recently visited Chicago). But I did leave Brazil, and after traveling and living in different places, I ended up living in the United States. My love for American music was also a strong pull factor. I grew up listening to American music. At home, my father listened to country and western (e.g., Jim Reeves, Hank Williams), Dixieland, traditional American folk, or skiffle music (i.e., Lonnie Donegan, who based his music on Leadbelly and Americana). In addition, I eventually became a blues aficionado and was infatuated with Americana.

However, after staying a month in New York City, I traveled to London. I was looking for a temporary job. After spending time washing dishes at a restaurant kitchen and busking with my guitar at several of London's underground stations, I answered an advertisement in a London newspaper. I signed up to work as a paid temporary agricultural worker in the Golan Heights, on the Syrian border—to work as a farmhand on a Moshav (an Israeli cooperative farm). I lived in Maale Gamla, a small community of farmers. Moshavs frequently recruit non-Jewish travelers, such as myself, to work as temporary agricultural workers. After my arrival, I cut my long hair in Tiberius. I also worked on date trees along the Jordanian border, picked bananas in the mud, or grapefruit, but I worked mostly in the vineyards on the hills overlooking the Sea of Galilee (Kinneret). I woke up at four in the morning every day, and worked ten to fourteen hours a day, six days a week. I believed that I needed to experience a sense of redemption and ablution through hard toil, even if I knew it would be only temporary. There is nothing easy or fun about hard agricultural work, yet it provided me with a different reality. I had spent my life in Brazil with live-in maids, drivers, gardeners, and seamstresses at home who did everything for our family. The Golan Heights was as far away as I could get from anything I had ever known. I wanted to erase my personal history. Ironically, I am doing the exact opposite here.

After four months, I was able to save my hard-earned shekels from a few months of grueling work, and I traveled southwards and worked on construction for a while in Eilat. I ventured into the Sinai Desert and lived in a Bedouin camp for almost a month, and then I went to Cairo for the first time. I returned to West London, where my aunt and uncle lived. I answered another newspaper ad, and secured another temporary job. This time, as a bartender in Scotland at a hotel owned by the Duke of Athol in the Central Scottish Highlands, at the Tilt Hotel in a village named Blair Athol. I worked there for the entire summer and winter. I learned how to pronounce the names of over 60 different single malts as well as appreciate the banter over the bar counter with local patrons. Traveling to the Middle East and then working in Scotland, I felt as if I had experienced another cycle of rebirth; as I explored place at the same time, I also was discovering *self*. As Richard W. Wilkie states: "Every person has feelings and attitudes about places that they have known and experienced . . . many times our relationship to a certain place is entirely unique to a particular time or series of events." (2003, 29).

I returned to the Middle East later as an undergraduate (through the Study Abroad program at Northeastern University), and was a student at the American University of Cairo for one semester and the University of Tel-Aviv for another. I was once again drawn to Cairo, where my grandmother was born and raised. I rented a furnished apartment with elongated doors and huge windows in the neighborhood of Zamalek, a block away from the Nile River and close to the Zamalek Jockey Club. I could walk or take the bus to campus, which was located beside a bustling Tahrir Square. The ubiquitous Cairene sounds of honking cars with the call to prayer blaring from loudspeakers were haunting and puzzling to me, as I would often notice centuries of dissonance living side by side: a donkey cart next to an automobile; or the smells of jasmine perfume and ancient incense mixed in with smog. I imagined what it must have been like when my grandmother lived on Soliman Pasha Street in a cosmopolitan Cairo decades ago.

I completed my undergraduate degree at Northeastern University (Boston, Massachusetts), and eventually settled in western Massachusetts with my then wife. And lastly, I departed for Baltimore to begin yet again a new chapter in my life, a new dimension of my autobiogeography.

TRANSNATIONAL BACKGROUNDS

I was born in Rio de Janeiro, Brazil, in 1967, and grew up bilingual (Portuguese and English). I was born and brought up Catholic (non-practicing) like my mother, Josette Hughes, who is a *carioca* too (a Rio native), and who also taught me how to surf and swim the waves of the Atlantic Ocean. She was born and raised in Rio, the daughter of an Englishman, Edward Hughes, and an Egyptian-born mother, Yolanda Imbruglia, who had arrived in Brazil in 1926 (of Sicilian and Italian heritage). Josette grew up in Petrópolis (named after Brazil's second monarch, Dom Pedro II, "city of Peter"), a small city located outside of Rio; at that time, considered the residence of Rio's crème de la crème, although her family was middle class. The climate was less humid than Rio because of its higher elevation, and it was favored by Brazil's late nineteenth-century aristocracy. She attended a French school there, run by French nuns (Colégio Sion), and became fluent in three languages (French, Portuguese, and English).

My maternal great-grandparents, Alessandro and Egizia Imbruglia (whose families were from Sicily and Lucca, Italy, respectively), were living in Cairo at the time my maternal grandmother, Yolanda Imbruglia, was born in 1908. Later, her family moved to Rio, and she spent her early adulthood residing in one of the first houses built on Ipanema Beach, Rio. I came to discover that the funds raised to build the house in Ipanema had been wired from Memphis, Tennessee, in the 1920s.

The world was globalized before the term "globalization" was a trend. The Imbruglias held familial and financial ties in Sicily, Lucca, Cairo, and Rio, and also the United States. These ties formed a cross-Atlantic, cross-Mediterranean triangle of sorts, enabling the movement of funds, within a transnational context. The presence of German, Italian, Jewish, French, and British communities, schools, hospitals, and a busy international community added a cosmopolitan flair, and Cairo's cultural and multi-religious diversity, as well as its diverse cosmopolitan, diplomatic, business, and trading hub, are facets that are likely to have easily translated into the Imbruglia family's new geographical knowledge of Rio, when they eventually moved there. These same transnational landscapes—cultural, religious, diplomatic, and ethnic communities—existed in Rio at the time of Yolanda's eventual move there. Furthermore, like Cairo, Rio was the nation's capital city at that time. Despite being born in Egypt, all of the Imbruglias had Italian passports and were staunchly Roman Catholic, and spoke five languages fluently. Their transnational experience would continue as they moved to Rio.

Alessandro Imbruglia, my maternal great-grandfather, was from Milazzo, Sicily, and had taken a job as one of the foreign-contracted engineers to work on the construction of the Aswan Dam in southern Egypt. He raised his five children in Cairo: Natalina. Maria Grazia, Yolanda (my grandmother), Alfonso, and Salvatore. At that time he amassed significant financial debt due to a gambling habit, losing most of his life savings. He decided to leave Cairo with his family for Rio. Egizia's brother, Silvio Nardi (the Nardi branch of my maternal family were from Lucca, Italy), owned a very profitable "Saloon" in Memphis at that time (Egizia's maiden name was Egizia Nardi Marucci). I have possession of a letter written by Maria Grazie (Silvio Nardi's great niece, and Yolanda's sister) about Nardi's newspaper obituary in 1913. It reads:

> He started a little grocery and saloon in South Memphis. He prospered and saved, sold and re-invested. He was one of the humble developers

> of Memphis. He accumulated a fortune, said to be worth now a quarter of a million dollars. Only a wife survives him to inherit his fortune.

Apparently, Silvio Nardi had wired the funds from Memphis to his sister, Egizia, in Cairo, to help build the new house on Ipanema Beach in Rio. Half a century later, in the 1970s, I remember as a child visiting that same house on Ipanema where my great-aunt, Maria Grazie ("Tia Mary"), still lived by herself, with no children of her own. Like Yolanda Imbruglia, her sister, she still wore jewelry (i.e., bangles) that she had brought with her from Cairo, and told us stories about their old home when her family lived in Cairo on Soliman Pasha Street, not too far away from Tahrir Square. I also visited that street when I lived in Cairo in the mid-1990s, while I was an undergraduate student for one semester at the American University of Cairo.

The Imbruglia house in Ipanema was one of the last remaining chic bastions of Rio's Belle Époque still standing, and later, in the 1980s, one of the last residential houses to be sold to a development company and then torn down. Today, in its place, is a gargantuan skyscraper. In front of the building remains an old single tree from that period.

My maternal grandfather, Edward Hughes (born in England in 1895), remained in Rio until he died in 1986, and was buried in the British Protestant Cemetery of Gamboa, in Rio. During World War I (1914–1918) he had been detained as a civilian prisoner of war at Ruhleben Camp (Berlin), as a British citizen working as a clerk in Germany at the time. After the war was over he was released from Germany and returned to England. He then took a job as a clerk and moved to Brazil. The disparity between the wretched barracks at Ruhleben and Rio at the time for him must have seemed starkly and unimaginably different. His arrival to tropical, lush Rio must have felt heavenly, especially after spending four hard years at Ruhleben Camp. He married Yolanda in Rio, and they had two daughters, Josette (my mother) and Jane (my late aunt). My father, Derrick Marcus, who was born in Cork, Ireland, in 1937, a Protestant, moved to London when he was ten years old, and he left for Brazil in 1963. My father, a chartered accountant, was contracted as an accountant to work for a British company in Rio. Brazil's national soccer team during the 1950s and 1960s contributed to my father's fascination with Brazil, with world-famous players such as Pelé, Garrincha, and Gilmar. He became involved in the Anglo-Brazilian and British community activities in Rio, and my mother was working as a primary-school teacher at the British school in Rio, where they eventually met and married. He remains actively involved with British community charity work in Brazil, and in 2002 was awarded an OBE (Order of the British Empire) in acknowledgment for his work. These familial ties and vignettes offer a context to the transnational scope of my autobiogeography. I will now explore my own perceptions of place-based experiences.

DISPLACEMENT

The fragmentation between a *carioca* (a native of Rio) and a *paulista* (São Paulo native) appeared distinctly different categories, where each carried strong archetypal personalities of place. For example, the *carioca* stereotype is of one who indulges in carnaval, beaches, pleasure, and leisure, while the *paulista* is known for his/her industrialism and hard work yet lacks any sense of swagger, in the eyes of the *carioca*. My family moved back and forth between São Paulo and Rio because my of father's job—where he accepted the office transfers. I felt a deep connection with Rio, a genius loci, since Rio is my birthplace and my mother's too—a place where my great-grandparents built a house and lived, and where my grandparents as well as my parents met and married.

My first home in Rio (they were all apartments in Rio) was on Rua Ataúlfo de Paiva in Rio's section of Leblon; my second home on Rua Gomes Carneiro, on Ipanema Beach; and, my third home on Delfim Moreira, Leblon. Then my family moved to São Paulo when I was four years old—my fourth home. When I was six years old, my family moved back to Rio, and my fifth home was on Copacabana Beach, an oceanfront apartment on the sixth floor, on Avenida Atlântica, a few blocks away from the famed Copacabana Palace Hotel.

It is commonplace to find maids working in Brazilian homes and not exclusive to middle or high social classes. Part of my father's job perks allowed my family to hire a live-in maid, who slept in her own room and who took care of the cooking and regular housecleaning; a daily maid, who took care of the dusting and cleaning chores; a gardener; a driver; and a seamstress. The maids who worked at our homes typically listened to transistor radios in the kitchen, and I remember hearing Roberto Carlos, of course, and also European hit songs in the late 1960s/early 1970s, such as "Aline" by French singer Christophe, or "Tornerò" by the Italian band I Santo California. I came to learn that my American peers were not familiar with these songs. The only other place I ever heard those songs played was at a restaurant in Cairo, much later in the mid-1990s. Our maids would also watch Brazilian telenovelas such as *Fogo Sobre Terra* or *Selva de Pedra*, and as a small boy, I would keep them company in the evenings, joining them as they watched on black-and-white television. Later, in the late 1970s, when my family bought a color television, I would watch reruns of popular American television shows which ranged from popular cartoons to varied popular shows of that time, such as *Charlie's Angels, Kung Fu, Bewitched, Batman, Bonanza,* and so on. And due to the largest Japanese community outside of Japan living in Brazil, we were also exposed to Japanese cartoons such as *Speed Racer*, or shows such as *Ultra Seven*, on Brazilian television. Thursday nights featured the "Spaghetti Western" genre films on television (a show called *Bang, Bang a Italiana*), which included the legendary Clint Eastwood films. I watched them religiously. All the foreign television shows were dubbed into Portuguese (no subtitles). All these examples here provide another transnational context and circumstance. I want to explore further the autobiogeographical context below.

I remember the utter humidity in Rio. Even before eight in the morning when I was getting ready for school, I would break out in a sweat. I attended the British school of Rio at that time. In the afternoons after school, I can remember walking across Avenida Atlântica over to Copacabana beach. At that time, crime had not yet reached the ubiquitous levels that it has today, and most crimes were not as violent. My father would often be away traveling for weeks at a time due to his job, so I spent a great deal of time with my mother and younger sister. My mother taught me how to respect the ocean and not to fear it. Therefore, swimming and surfing provided that rare solace in my childhood, one of the very few activities where I felt very comfortable and safe, since I was alone almost all of the time. I felt a deep sense of social detachment and alienation, even at the age of six years.

I can still remember the smell of the soft afternoon ocean mist, as I would be comforted by a light breeze and waves lapping on my feet. I would always be able to see Rio's granite peaks jutting in the backdrop that surrounded me and seemed to shelter me as if they were the arms of my deceased forefathers/foremothers looking on. The physical landscapes gave me a sense of location, a genius loci; I knew where I was located based on Rio's visible iconic landmarks, ranging from the peaks of the *Pão de Açucar* ("Sugarloaf"), *Pedra Bonita* ("beautiful peak"), *Dois Irmãos* ("Two Brothers"), and *Corcovado/Cristo Redendor* (the iconic "Christ Redeemer" statue on Corcovado). Even when I felt lost, I could find my bearings.

In 1976, our family moved for the last time, when we returned to São Paulo—where my parents still reside. My mother was displeased that we needed to leave Rio. My sense of alienation only increased. Now we lived in a large airy house in São Paulo in the highly coveted residential upper-middle-class neighborhood of Brooklin (spelled with an "i"). Many of the houses in that neighborhood were originally built by German emigrés in the 1940s. Former Nazi physician Josef Mengele was purportedly harbored by fellow emigrés and German descendants in that neighborhood, with its strong German loyalties and ties (in a similar way, Adolf Eichmann had been harbored and protected by the German community in the outskirts of Buenos Aires).

I owned a ten-speed bicycle in 1977 (a green Caloi 10). And in the same way that the sea provided a refuge in Rio, that bicycle provided me with a mode to explore the city as a means to escape and encounter reality at the same time. I remember when part of a major city artery along the Pinheiros River, named the *marginal,* was still under construction and still shut off to traffic, I would sneak in, and ride miles of empty asphalt on that section of the *marginal* alone. That particular *marginal* section would connect to what is now known as the Morumbi Bridge. Back then, the artery was a major connection to Interlagos. The Morumbi Shopping Centre had not yet been built. The only other people I would see during my ride were random groups of Gypsy/Roma living in camps on the side of the river. I would often ride into other regions too, by rogues and rum drinkers, *cachaçeiros* in the streets, and shanty towns, *favelas*. I rode everywhere it seemed, without fear. Every day of the week, whether it was warm or cold, it did not seem to matter as I always wore shorts outdoors, and despite the fact that the Tropic of Capricorn runs right through the city, winters in São Paulo can be bleak, rainy, damp, and cold (houses did not have any indoor heating). I was acutely aware of the often dangerous, polluted, and urban geographies that surrounded me even if I was sheltered from those realities, I was well attuned to the extreme levels of poverty and wealth living side by side.

Later, when I was fourteen years old and wore flannel shirts and had longer hair, I had a gun pointed at my head for the first time by the military police (the military dictatorship in Brazil lasted from 1964 to 1985). I was returning to my house from the cinema, walking back home with a school friend a few blocks on Rua Joaquim Nabuco, and the military police van, *camburão,* pulled up with their guns, pointed one at my head, and asked us, "Where we are you going? Where are the drugs?" We had no drugs, just our national identity cards, *carteira de identidade,* but we looked subversive perhaps, and they treated us as such. They eventually let us walk away. Institutionalized violence in Brazil was rampant and still is (i.e., military or civil police, civilians, and criminals alike). One of the reasons I left Brazil was that I no longer wanted to live with the daily threat of violence as a mechanism to resolve domestic or routine public disputes. Ironically, I was never robbed or mugged in Brazil. I needed to go to Europe for that to happen. At the age of seventeen, while I traveled with a friend in Europe, I was mugged at knife-point in an Amsterdam alleyway.

BILINGUAL SCHOOLING, TRANSNATIONAL FRIENDS

In São Paulo, I attended a private British school called St. Paul's School and, later, an American high school, *Escola Graduada,* also known as "the American Graded School"—two of the most prestigious bilingual schools in Latin America. In both schools, classes were taught in both English and Portuguese. We took classes (in English), for example, in mathematics, biology, chemistry, physics, geography, history, English literature, and British poetry and read for example, *Julius Caesar, Dr. Jekyll and Mr. Hyde,* and *Pygmalion,* among others. At the same time, we also took similar classes taught in Portuguese, such as history of Brazil or geography of

Brazil. At Graded, I read the Brazilian and Portuguese literary and poetry greats, such as Luís de Camões, Gonçalves Dias, Monteiro Lobato, Euclídes da Cunha, and Mário de Andrade.

Most of the students of both schools at that time were either wealthy Brazilians or children of expats who were working on contract in Brazil, working for multinational companies (including Swedes, Finns, British, Japanese, Americans, and Italians). In São Paulo the French had their own school, Pasteur, the Swiss, Colégio Suiço-Brasileiro, the Germans, Porto Seguro. The Americans had their own private schools: Chapel School, Graded, and Pan American Christian Academy, PACA (just outside of the city of São Paulo, in Campinas).

It is well known that the terms "ex-patriate," "ex-pat," (or even "sojourner") are commonly used to describe the British in Brazil, but never, "immigrant." The term "immigrant" still carries a social stigma among the British in Brazil (see Marcus 2011a), the British never referred to themselves as immigrants. The most common form of fraternizing within the British "ex-pat" community in Brazil was time spent in social clubs (i.e., clubs dubbed "athletic clubs" such as S.P.A.C., Rio Cricket Club, Paissandú Club), and sending their children to British schools in Brazil (i.e., The British School of Rio, *Escola Britânica,* or St. Paul's School). The student population at St. Paul's School was not all British-born, in fact, they were diverse and mostly Anglo-Brazilians or Brazilians. There were also many Jewish students as well as Brazilian-born descendants of Middle Eastern emigrés, mostly from Lebanon and Syria.

Every Monday morning, students met in the school's assembly hall (past the photograph portrait of Queen Elizabeth II and the Union Jack flag). We all would be waiting, sitting down until the diminutive school headmaster, Mr. Ross, would swiftly walk with his formal robe fluttering behind, through the quiet rows of students dressed in student uniforms (i.e., blue blazers for boys, red blazers for girls; white shirts, school tie, black shoes, grey socks, and grey pants or skirts for girls). Students would immediately stand up in reverence as he walked, in a quiet, solemn procession. Mr. Ross would then commence the assembly meeting at the podium on stage by making the weekly announcements in his staunchly unemotional Scottish accent. The local Anglican vicar (also British) would then provide the assembly with a sermon, and ending the meeting we would sing a hymn or two, such as "Onward Christian Soldiers," as Mrs. Tuthill played the piano. On special occasions we would sing the British national anthem, "God Save the Queen." However, I am not affiliated with the Anglican Church. I was born Catholic, my Brazilian mother is Catholic, and I was born and raised in the staunchly Catholic country of Brazil, the world's largest Catholic country. Nonetheless, my upbringing was tinted by a cold, insular, detached British tinge (albeit an archaic circa-1930s Britain).

Later, I attended the American Graded (1982–1985), where I graduated with a high school degree accredited in both Brazilian and U.S. educational systems. Exposure to the U.S. educational system came with a dramatic change for me personally and academically. In this school there were no school uniform requirements, and girls wore tight blue jeans, lipstick, and makeup, and chewed gum, and the general mise-en-scène became somewhat of a novel fascination to me. It felt to me at that time as if I had switched from watching black-and-white to color television. I took an interest in poetry, writing, and English literature, and the school literary magazine even published a few of my poems and short stories. I did well academically for the first time in my life. I favored the broad and welcoming U.S. educational system—and that also contributed to my decision to live permanently in the United States.

BECOMING A PROFESSIONAL GEOGRAPHER

My parents had taken me along with them during their travels around the world since I was nine months old and, thereafter, took me with them to London every summer until I was probably eighteen or so. Those trips are likely contributing preludes to my later viewpoints as a geographer. My father would take "home-leave" (as they called it) every year. He was employed as an accountant at a British-based multi-national company, based in London (he was an employee in Brazil since 1963)—one of the perks of his job. Instead of returning to London, my parents would take the "long journey" back to London which included visiting Australia, South Africa, Morocco, and Mexico, to name a few countries. We would often return from England to Brazil by ship, in the 1960s/1970s on the Blue Star Line (from Southampton to Rio). I had traveled to every continent except Antarctica before the age of sixteen. Later, I traveled on my own throughout many countries and I would often save money working in one place and travel to the next.

Hence, I had already become a geographer but did not really know it until much later. I met my academic mentor, Richard Wilkie, at the University of Massachusetts, Amherst. I also was fortunate that I had the encouragement of my wife at the time, Patty, and the convincing boost of another graduate student, George Roberson, who had talked to my wife to encourage me to join the geography program. At that time I was working at *The Recorder*, a local newspaper in Greenfield, Massachusetts. As I have looked at my familial transnational ties and background, I will briefly reflect on my influences as a professional geographer and my academic genealogy. In 1968, Richard Wilkie, virtually single-handedly, started the geography program at the University of Massachusetts, Amherst, where he is now Professor Emeritus. Wilkie helped direct my professional perspectives as a geographer, as he opened doors that helped to shape my own geographical visions as a graduate student, and reinstilled and restored my sense of intellectual self-esteem. More importantly, it was through my passion for the discipline of geography that I came to better understand and sort out the complexities in my own personal life. Its disciplinary lenses were as rigorous as they were broad and flexible, and I felt I could finally explore various academic research interests of mine.

Several graduate courses transformed my way of thinking. However, "Visual, Graphic and Spatial Thinking," taught by Wilkie, was one of those fundamental courses where I learned how to project data and theories I had envisioned onto a visual or graphic or map. Wilkie's pedagogical perspectives were inspiring, and I experienced his excitement through his teaching and, by proxy, my own excitement of learning, something I completely lacked decades ago as a schoolboy. Spatial thinking helped me organize my dissertation research ideas, as it has continued to do to this day. In that same geography department, I also took graduate courses with Piper Gaubatz, Stan Stevens, and Julie Graham. I also took two instrumental graduate classes in the history department with Mary Wilson—who taught me how to be "ruthless" as a writer—and with Jane Rausch.

Wilkie attended the University of Washington, and in 1950s was enrolled in undergraduate courses taught by Thorsten Haagerstrand and Dauril Alden (who sat on his dissertation committee), among others. He was a personal friend of the late Wilbur Zelinsky (at one point he was Carl Sauer's oldest living student from the University of California, Berkeley)—I had met him a few times at Wilkie's home. It was Zelinsky who introduced Wilkie to his wife, Jane, a retired sociology professor. Wilkie's graduate school colleagues, such as Waldo Tobler, and his mentor Richard Morrill, also helped shape his scholarship and teaching methods and wide-ranging approaches—which, in turn, indirectly also affected my own perspectives in geographical thinking. He had conducted his doctoral research in Argentina, tracing the trajectories of the Volga-Deutsch population outside of Buenos Aires.

Long before the term "transnational" was in vogue, Wilkie's cutting-edge research looked at various dimensions that were, in fact, transnational *par excellence*. He had examined the interrelated ties and loyalties of generations of migration patterns from and to Germany, Russia to Argentina. He was also an undergraduate student at Mexico City College, together with other students who were later to become well-known academics. He was a graduate student at the University of Washington, which at the time, was considered to be the epicenter of geography's so-called "quantitative revolution." Ironically, later in his life, his theoretical perspectives were shaped by qualitative methods, humanistic geography, spirit of place, and sense of place (as well as quantitative perspectives).

As well as a geographer, he became a respected and award-winning photographer. Today Dick continues to be a mentor and a great friend.

I conducted my doctoral dissertation fieldwork research under Wilkie's supervision and also with helpful insights from Joya Misra, a sociologist, who also sat on my dissertation committee, for which I am grateful.

My doctoral research focused on Brazilian immigrants in the United States. I was greatly influenced by the work of Maxine L. Margolis and her groundbreaking research with Brazilian immigrants in the United States (1994, 2013). I had contacted Margolis before I began graduate work, and not only did she provide me with supportive insights during my research work, but also continued to provide me with an inspiring context for my later publications about Brazilian immigration. I successfully completed my MA and my PhD degrees from the geography program at the University of Massachusetts, Amherst. Moving forward, I will look at my recent trajectory to Baltimore.

BALTIMORE

I accepted a job offer as assistant professor at the Department of Geography and Environmental Planning at Towson University in 2008. My colleagues in the department as well as the College of Liberal Arts were, and have always been, graciously welcoming to me and since then, I was awarded tenure promoted to associate professor. The *experiencing of place* often means the *discovering of place and self* at the same time. Accordingly, Richard W. Wilkie stated: "For many people the discovery of special places through one's explorations can be nearly as powerful as revisiting one's own pantheon of sacred places" (2003, 31). In this vein, Baltimore has become one of those few places in my pantheon of sacred places. When I moved to Baltimore, I remember how I felt a strange sense of "belonging."

I wanted to know and experience Baltimore as much as possible, as a geographer who investigates place. I explored this city with its torrid warmth, quirkiness, and grit. I also participated in unorthodox interests which offered different insights into the city and Baltimoreans. Even when middle-age angst reached me, I was fortunate enough to have met the inspiring Towson University boxing club coach, Herb Hollidge, who invited me to join a local boxing club, Charm City Boxing, training with other boxers and coaches (such as David Sewell), which gave me a sense of perseverance and assuaged that angst. I signed up for a poetry workshop headed by local Baltimorean poet Shirley Brewer and started writing poetry again. I played lead guitar for the Billy Thomson Band around Fells Point. I created and organized a successful event named "Geopoetics" at Towson University with Baltimorean poets, such as Clarinda Harris, Shirley Brewer, Michael Ratcliff, Barbara Morison, and Leslie Harrison, and geographer Jon Hessler—where we discussed the poetics of place. In summary, I "went native"—as they say in anthropology.

I felt that I had fit in discretely and that this city had adopted me openly. Yi Fu Tuan's concept of *Topophilia* (1974) underlines how we can develop emotional attachments to place(s); that is, attachments which are powerful to each individual who experiences it (from *topos*=place + *philos*=love: "love of place"). For example, as Baltimorean author Rafael Alvarez (also one of the writers for the HBO series *The Wire*) said it best in his passionate love letter to the city of Baltimore, which resonated with me, he begins: "My Beloved Crabtown, I could live anywhere in the world—in storybook capitals or pastoral hamlets greatly desired—but, God help me, I choose to cohabitate with you" (2015, 439). Likewise, in my case, of all the places I lived and experienced, from Rio and Cairo to London, Blair Athol, and of all bending roads in betwixt and in between, the one that led to Baltimore was the most eccentric and alluring. Alvarez ends his letter poignantly: "I know that anywhere, at any moment, I may be kissing your cheek for the last time" (2015, 441). Similarly to Alvarez's sentiments, I also feel emotionally connected with a strange, uncomfortable sense of impermanence to Baltimore—as if I had to say "good bye" to a woman I love and whom I would never see again.

Where is "my home"? This is my home now.

Anne Buttimer had aptly stated: "To dwell implies more than to inhabit, to cultivate, or to organize space. It means to live in a manner which is attuned to the rhythms of nature, to see one's life as anchored in human history" (1976, 277). And I felt attuned to the rhythms of this city where I now dwell. I breathed in Baltimore only to discover this place with all the upheavals of human sentiments. From dive taverns, pubs, the famed iconic Domino Sugar red neon sign, to "Natty Boh's" local beer, the Orioles Baseball team, the Baltimore Ravens football team, the Cat's Eye Pub, Leadbetter's Tavern, the "Night of 100 Elvises" event, Belvedere Square, Roland Park, Hampden, and the warm cobblestone streets of Fells Point—I stayed here. For the first time in my life, I did not leave a place (yet now, I am the one who begs Baltimore: *ne me quittes pas*).

I remember when I was first embraced by a warm, breezy Baltimorean September night, which took hold of me with all its grip. Along the misty streets, I could smell fresh-baked bread in the air from a local bakery. I also remember when I stood in a whirlwind of patrons watching a live band at the Cat's Eye Pub in Fells Point. Amidst musicians, dancers, doctors, accountants, poets, writers, nurses, junkies, professors, teachers, tipplers, sinners, saints, hoodlums, hustlers, and the dregs of skid row—they all found a sense of place in that perfectly dystopian world. No matter how dysfunctional their lives, they all moved around in sync and interacted functionally for a few hours, uninhibited with unrestrained abandon. Nobody's last name, nationality, or profession mattered. Like the smell of amber sandalwood, Baltimore had seduced me with her charm.

Most people do not typically aspire to "end up" in Baltimore, especially with well-known public attention given to the imageries of the recent Baltimore riots or scenes from *The Wire* episodes. Yet I embraced it with all its grit and unassuming personality, which I found charmingly refreshing. And the *personality of place* is reified by the fact that Baltimoreans are one of the friendliest people I have ever met—they always seem genuinely interested in talking to others, especially out-of-towners. I explored these new physical and cultural landscapes, as well as new emotional landscapes.

I noticed how much I appreciated those amiable Baltimoreans and their unpretentious *Baltimorese* accent (e.g., "ambue-lance," "owe's," and "ohn's")—since I arrived in this city, I have felt the same affection. Charm City is indeed an apropos nickname for this uninhibited, oddball place, with her unassuming, nostalgic personality (i.e., Baltimore *is* nostalgia). I miss her terribly, but I never left.

CONCLUSION: BURIED MEMORIES OF *SELF*

All too often, concepts of ancestral homeland and nationhood are reiterated in the social sciences or cultural studies literature. "What is your identity, national, cultural, religious, ethnic?" Mine is fragmented, hybridized, and displaced, and transnationalism is the appropriate theoretical backdrop to help explain this phenomena. One of my students asked me during the last day in one of my classes: "Dr. Marcus, just what is your ethnicity?" Despite the fact she knew I was born and raised in Brazil, she was still confused. I also cringe when I am asked: "*What* are you?" Inevitably the questions and comments which follow are predictable, such as: "But you don't look Brazilian" or "You speak English very well." I also need to keep a set of stock answers such as "I do not speak Spanish," or that the extreme easternmost point of Brazil is physically closer to the westernmost point of Africa than it is to the extreme easternmost point of Peru.

My own cultural and social perspectives were shaped within a periphery of sorts in Brazil, and later as a young adult outside the comfortable boundaries of family life. I lived in a counter-culture periphery. In this regard, my own *Brazilianness* within Brazil, even before I left Brazil, was already displaced and marginalized. For example, I am the antithesis of the Brazilian joie-de-vivre, archetypical spirit: I do not partake in carnaval celebrations, or dance to samba, nor am I a practicing Catholic, or an avid fan of soccer. Does that make me less Brazilian? I felt I was doomed to live my life as a perpetual outsider, transgressing on multi-scalar levels (family unit, societal, national, and international). Therefore, I never met others' expectations of what I *ought to be*.

In order for me to be perceived as an authentic Brazilian, outside of Brazil, should I need to speak English with a thick accent? Or eat *churrasco* (Brazilian-style barbeque), listen to *sertanejo* music (a rural genre of Brazilian music)? Am I less Brazilian because I do not "look the part," nor do I "behave/speak" like the stereotypical Brazilian? Would I have more academic clout (authenticity) as a Brazilian if I spoke with a thick accent? Would that *legitimize* me as a "typical Brazilian" in academic and public spheres? These questions challenged me to think carefully as I designed and carried out my own research on Brazilian immigrants. The fact is that I was *Americanized* before I left Brazil, like many other Brazilians—a product of a globalized world, consuming U.S. pop culture and music; however, and more importantly, my family background before I was even born was transnational. This discussion had also helped me with predicaments of my own research design in the past as well as the understandings of nationhood, cultural norms, and ultimately about the disciplinary practices of geography (human movements and perceptions) and transnationalism.

I *ought* to belong to an exclusive disciplinary practice, cultural or ethnic category, in order to calm the fears of those who seek an easily digestible identity at hand even in a global age. However, my transnational transgressions are subtle, almost invisible, yet fragmented and permanent. Transnational geographers render effective the richness of multiplicity, stemming from the assortment of geographical experiences and imaginations, by transcending boundaries through hybridity and displacement. I am writing now to understand about *self*, while others are busy making their own memories—most of mine remain buried. And like most of us who have left a homeland forever, we know that we have left some sort of unfinished business behind.

REFERENCES

Alvarez, R. 2015. *Crabtown U.S.A.: Essays and Observations*. Cibolo: Texas: Perpetual Motion Machine Publishing.

Basch, L., N. G. Schiller, and C. S. Blanc. 1995. From Immigrant to Transmigrant: Theorizing Transnational Migration. *Anthropological Quartely* (1): 48–63.

Buttimer, A. 1976. Grasping the Dynamism of Lifeworld. *Annals of the Association of American Geographers* 66(2): 277–292.

Castles, S., and M. J. Miller. 2003. *The Age of Migration*. Third Edition. New York: The Guildford Press.

Entrikin, J. N. 1976. Contemporary Humanism in Geography. *Annals of the Association of American Geographers* 66(4): 615–632.

Friedman, M. and S. Schultermandl (Eds.). 2011. *Growing Up Transnational: Identity and Kinship in a Global Era*. Toronto, Canada: University of Toronto Press.

Hay, I. (ed.). 2000. *Qualitative Research Methods in Human Geography*. Oxford, United Kingdom: Oxford University Press.

Hoggart, K., L. Lees, and A. Davies. 2002. *Researching Human Geography*. London, United Kingdom: Arnold Publishers.

Levitt, P. 2001. *The Transnational Villagers*. Los Angeles and Berkeley: University of California Press.

———. 2004. Transnational Migrants: When "Home" Means More Than One Country, *Migration Information Source*. [http://www.migrationinformation.org/Feature/print.cfm?ID=261].

———. 2007. *God Needs No Passport. Immigrants and the Changing American Religious Landscape*. New York: The New Press.

Marcus, A. P. 2015. Convenient Liaisons: Brazilian Immigration/Emigration and the Spatial-relationships of Religious Networks, *Space Populations Societies* (2014-2-3) 2–12.

———. 2013a. Sex, Color, and Geography: Racialized Relations in Brazil and its Predicaments. *Annals of the Association of American Geographers* 103(5): 1282–1299.

———. 2013b. Global moves from and to Brazil. In *Careers without Borders: Critical Perspectives*, 93–106, edited by Y. Baruch and C. Reis, Oxford, UK: Routledge.

———. 2011a.Transnational Rio de Janeiro: (Re)Visiting Geographical Experiences. In *Growing Up Transnational: Identity and Kinship in a Global Era*, pp. 21–35, edited by May Friedman and Silvia Schultermandl. Toronto, Canada: University of Toronto Press.

———. 2011b. *Towards Rethinking Brazil: A Thematic and Regional Approach*. NJ: John Wiley & Sons, Inc.

———. 2011c. Rethinking Brazil's Place within Latin Americanist Geography. *Journal of Latin American Geography* 10(1): 129–147.

———. 2011d. Experiencing Ethnic Economies. *Journal of Immigrant and Refugee Studies* 9:1–24.

———. 2010. Back to Goiás and Minas Gerais: Returnees, Geographical Imaginations and its Discontents. *Revista Tempo e Argumento* 2(2): 121–134.

———. 2009a. Brazilian Immigration to the United States and the Geographical Imagination. *Geographical Review* 99 (4): 481–498.

———. 2009b. (Re)Creating Places and Spaces in Two Countries: Brazilian Transnational Migration Processes. *Journal of Cultural Geography* (26)2: 173–198.

Margolis, M. L., 1994. *Little Brazil: An Ethnography of Brazilian Immigrants in New York City*. Princeton, New Jersey: Princeton University.

———. 1995. Brazilians and the 1990 United States Census: Immigrants, Ethnicity, and The Undercount. *Human Organization* 54: 52–59.

———. 1998. *An Invisible Minority: Brazilians in New York City*. New York: Simon and Schuster.

Tuan, Y. 1974. *Topophilia: A Study of Environmental Perception, Attitudes, and Values*. New Jersey, NJ: Prentice-Hall.

———. 1977. *Space and Place: The Perspective of Experience*. Minneapolis, Minnesota: University of Minnesota Press.

Tsuda, T. 1999. The Permanence of "Temporary" Migration: The "Structural Embednessness" of Japanese-Brazilian Immigrant Workers in Japan. *The Journal of Asian Studies* 58 (3): 687–722.

Wilkie, R. W. 2003. Sense of Place and Selected Conceptual Approaches to Place. *Journal of the American Institute of Architecture Students*. 55(1): 29–31.

TWO

Enjoying the Best of Two Worlds or Torn Between Two Places?

Heike Alberts

I never thought that I would become a university professor in the United States and travel back and forth between two continents on a regular basis.[1] Likewise, I never imagined that my personal and professional lives would end up being closely intertwined, as a transnational migrant studying highly skilled migrants living transnational lives. When I first started studying migrants as an undergraduate student, I assumed that migration meant a permanent move from one place to another, with people essentially exchanging one life for another. Even when I became a migrant myself, initially there seemed to be a strong separation between my place of origin and the place I lived. Over the years, however, my own life became increasingly transnational at the same time as transnationalism became a key concept in the migration literature. Living and studying transnationalism simultaneously has enriched, but sometimes also complicated my life. I still cannot answer the question whether living a transnational life means being able to enjoy the best of two worlds or being torn between two places. Living a transnational life also has implications for how I define myself. In my mind, I am a *German in America* when I am in the United States, and a *German living in America* when I am in Germany.

FROM BERLIN TO WISCONSIN

I was born and raised in Berlin, Germany, during the Cold War when the city was divided by the Berlin Wall. My parents had both grown up in

Berlin as well, and experienced the end of World War II, the construction of the Berlin Wall in 1961, and the fall of the Berlin Wall in 1989 first-hand. My parents have always had a strong attachment to Berlin, and would never have considered moving abroad. In fact, my father has spent 78 years of his life in the house my grandfather had built in Berlin.

While my parents (and indeed the rest of my family apart from a distant aunt who migrated to Canada many years ago) do not have any experience of living abroad, for as long as I can remember, we always traveled. It was often said that Berliners are the world travel champions, as many Berliners left town as often as they could. Being surrounded by the Berlin Wall meant that we could not do any weekend trips, so we always had to go far away. As a result, from a young age I was used to spending time abroad on vacations (e.g., in the Netherlands or Switzerland).

I traveled frequently with my parents from an early age, but I was rather shy and afraid of being away from them as a child, to the point that I did not even want to go to a friend's house to play. For anybody who knows me now that must be a big surprise as I am now outgoing and travel all over the globe on my own. The turning point for me came in 1985 when I traveled to England with our high school orchestra; by 1988 I really wanted to go abroad for a longer time period. My first longer stay abroad was to spend three months in Canada with my aunt's friends. Three years later, after finishing high school, I lived in Paris, France, for six months, working as an au pair and taking classes both at a language institute and the Sorbonne.

In 1992 I began my undergraduate studies at the Free University of Berlin, as a geography major and a North American studies minor. As soon as I finished my undergraduate degree, I applied for an exchange scholarship to the United States. Somewhat to my surprise, I received a one-year scholarship to Western Michigan University for the 1994/1995 academic year. The exchange year was a wonderful experience for me. Despite living in a tatty apartment with hardly any furniture, I have very good memories of that time. Even though I was surprised by the amount of work demanded of American students and the never-ending stream of exams, readings, and assignments, I did not find the classes very demanding intellectually. I probably did not quite realize it then, but now I know that students go through a much more rigorous high school system in Germany, so that they usually have the necessary academic skills to succeed and expect to work hard. I found that the classes in the United States covered a lot of material, but were not terribly challenging in terms of skills or level of thinking required. Especially the exams, usually in multiple-choice format, were much easier than what I had encountered in my German university. Assignments were numerous, but easy to do. It was also interesting to me that the classes were not only taught differently, but were also different in content. I loved the regional geography classes that we did not have at home, and got to know another way to "do geography."

Even though we all had lots of work to do for our studies during the exchange year, it was easy to make friends among other international students and enjoy ourselves in our free time. Six of us purchased an old Oldsmobile, and since gas was 99 cents per gallon, we did many weekend trips within Michigan, traveled to New York City over Thanksgiving break, and drove all the way to Key West over Christmas. It was a wonderful time. At some point I fell in love with another student from Germany who studied Mechanical Engineering at Western Michigan University. He was a year ahead of me in his studies. Because things went so well for him, he contemplated staying in the United States to finish his degree at Western Michigan University and possibly working in the United States afterwards. At that time, I could easily imagine staying in the United States with him, and did not even really think about what a huge decision that would have been. Maybe I was naïve at the time, but being in love probably led me to not seriously consider the impact staying in the United States would have had on my family and friends, or to ponder the question of where I belong that has become such a central part of my life now.

In the end we both ended up returning to our respective German universities and finishing our degrees there. Our relationship eventually ended when he accepted a job in the United Kingdom and I was not quite ready to give up my career to move with him. I did not want to be the "trailing spouse" with no clear professional future. Furthermore, even though the U.K. was geographically much closer to Germany than the United States, to me it felt more distant, as I was not familiar with the culture and felt no connection to that country. Of course I often wonder how my life would have turned out had I stayed in the United States then or gone to the United Kingdom with him. However, I have never regretted my decision not to go with him. I knew that just being a wife would not have fulfilled me. However, several of my close female friends ended up engaging in "love migrations." Both women, despite having advanced university degrees, ended up giving up their academic careers and dedicating themselves to their families.

I spent another three months in the United States in 1996, doing an internship with Dr. Thomas Boswell in Miami. This internship was an eye-opener for me. Not only were all the people I worked with very welcoming, but I was taken seriously even though I was only a student. In Germany, the university system is very hierarchical, so as students we had little contact with our professors and, especially in the teaching-track that I was in, virtually no opportunity to work with a professor. During my internship, however, I was allowed to work on an atlas of Puerto Rico with Dr. Boswell and his colleague Dr. Cruz. I learned a huge amount assisting with a major project and working with people who took the time to teach me various skills. In hindsight, this was probably when the seed was planted for my academic career. First, however, I returned to the Free University of Berlin to complete my graduate degree in geography and North American studies with the goal of becoming a high school teacher.

All my life, I have only ever seriously considered one profession: being a teacher. Even when other children wanted to be firefighters and astronauts, I knew that I wanted to teach. My parents tell me that I always taught in some shape or form—whether it was teaching ice-skating to other children, tutoring classmates who struggled in school, or later giving flute and Spanish lessons. The only thing that changed over the years was the level at which I wanted to teach. My university studies qualified me for teaching Geography and English as a foreign language at the high school level. During the three obligatory teaching internships I had throughout my studies I taught at three different levels (elementary school, comprehensive school, and college-prep school) and it was clear that college-prep was where I was headed. However, when I finished my degree in 1998, only a few positions were available for teachers in Berlin, and somehow I was not quite done with academia, so I decided to pursue a doctoral degree (as did two of my cousins, so that probably rubbed off on me). Since my main regional interest was North America, and I had heard that PhD students were funded in North America, I applied to two PhD programs in the United States. I really had no idea what I was getting myself into, but when I was accepted by both universities with funding, the big decision was not whether I would go, but where I would go. Eventually I ended up in the PhD program at the University of Minnesota.

When I started my PhD program in Minnesota, I had no intention of permanently staying in the United States. I expected that I would get my PhD degree in the United States and return home. I was not quite sure what I would do afterwards, but I assumed that things would somehow fall into place, so I just concentrated on getting my degree. As we came closer to graduation, my graduate school friends started applying for academic jobs in the United States, so it was logical for me to do the same. Two days before my graduation in 2003, I accepted a one-year lecturer position at the University of Wisconsin Oshkosh. Two years later, I was offered a tenure-track position at the same university and have been there ever since.

Even though my original plan was to return to Germany after graduation, the "path of least resistance" was to follow a career in the United States. However, as I was approaching graduation, I started to think seriously about the question of whether to stay in the United States permanently (as opposed to just another year or two) or return to Germany. Now, more than a dozen years later, I still do not have an answer to that question, but I have spent a lot of time reflecting about how people make migration decisions, how my life has changed as a result of migration, and what it means to live outside of one's home country. Meanwhile, there were significant changes in communication technology which have changed the experience of living abroad and to what degree one can stay in touch with home.

RESEARCHING MIGRATION INTENTIONS

When I first started thinking about the question of whether to stay in the United States permanently or return to Germany and talked it over with my close friend Helen Hazen (who is from the U.K.), it became clear that this was not only a question of personal interest, but also an interesting research topic for somebody who specializes in international migration. Helen and I decided to carry out a study of which factors international students consider when they ponder the stay versus return question, how they weigh the different factors, and how this varies among different subgroups.

Carrying out this study was fascinating for us. We organized focus groups with students from six different countries to discuss these questions. Many things said in the focus groups were familiar to us—for example, the idea of being torn between two countries, the feeling of liking the United States yet not really feeling at home there, and the different understandings of friendship, so that we very much identified with our research participants. However, we also saw some differences. For example, neither one of us comes from a culture with strong family obligations, so for us the "parents factor" is more a question of *wanting to return* to be there for parents rather than *having to return* to fulfill expectations. We also both come from rich, democratic, well-run countries with good pay, health care systems, and benefits, so we do not have to worry about potentially returning to political oppression, economic hardships, or conflict. So for us, some factors that weigh heavily for international students from other countries are not really a concern, but there are still many factors to consider. The research made it clear that even though the details vary for individual students, the decision of whether to stay in the United States or return home is agonizing for many.

Now, more than ten years after conducting this initial study about international students, we still work in the field. We have published several academic articles about the migration intentions and experiences of international students and scholars, edited a book about academic migrants, and have plans for further studies on this topic. As a result, our personal and professional lives have become interwoven as we deal with the same questions in our personal lives as in our research.

STAYING IN TOUCH WITH HOME

When I came to the United States to spend an undergraduate exchange year at Western Michigan University (Kalamazoo) in 1994, communicating via email was already standard at U.S. universities, but only starting out at German universities. I distinctly remember hearing the word "email" for the first time during our orientation before going abroad, and could hardly believe it—could you really type a message on a computer, and it would appear on somebody else's computer screen a few seconds later? I did not even have a computer at the time (I wrote all my undergraduate papers on a typewriter), so this seemed extraordinary to me. In Kalamazoo I had the chance to try out email for the first time. Since I did not know anybody who had email in Germany, all I could do was to email back and forth with the friends I had in Kalamazoo. I clearly remember how much fun we had sending messages to each other while sitting in the same computer lab just yards from one another.

Communication with my friends and family in Germany was rather limited during my exchange year. Of course we wrote letters, which took about a week to travel across the Atlantic, and every two weeks or so my parents and I spoke on the phone. Phone calls were expensive, about $10 for 10 minutes or so, so they were infrequent and short. Apart from my parents, only my grandmother called every once in a while, so I did not know much about what was happening at home or in Germany more broadly. Two friends from high school came to visit at some point, but overall I was pretty much cut off from everything German for the entire year. I do not think that this really bothered me, because I saw the exchange year as an adventure.

After returning to Germany from my exchange year, I of course wanted to continue to use email. We had an open access computer lab on the Free University of Berlin campus where the computers were online, so about once a week I stood in line for anywhere between 20 minutes and an hour to be able to use email for half an hour or so. By the time I left for the United States in fall of 1998, I had had dial-up email at home for some time, and my parents had learned to use it as well, so when I moved to Minnesota, email became our primary means of communication. Most of my friends now had email as well, so while we still wrote snail mail letters on occasion, we all learned to appreciate the joys of instantaneous communication.

During the late 1990s phone calls were still relatively expensive, but this eventually changed. As I write this chapter, a phone call from Germany to the United States costs about one cent a minute, and is sometimes even cheaper than a call within the United States, so I can communicate with family and friends much more often. I am now much more informed about what happens in Germany, as we do not have to limit our calls to exchanging the most important information, but can talk extensively about anything that happened or that we find interesting. Communication became even better when Skype became available. It was exciting when digital photography made it possible to share photos within seconds via email, but to see another person live while talking is almost as if one is sitting in the same room. Conversations overall are now much more as if one was "there" rather than thousands of miles away. For example, I enjoy doing crafts in my free time. So when I finish a project, I can show it to my friends and family by holding it up to the computer camera. Or when I bought new furniture and rearranged my apartment, I could take friends in Germany on a tour by carrying my laptop with the camera around. The ease and cheapness of communication also makes it possible "to be there" for others when needed. For example, when my father was in and out of the hospital a few years ago, I could communicate with my mother several times a day to hear the news and comfort her. It was not quite the same as being there, but the next best thing. For me, technological changes have shrunk the distance in a very real way, as I can communicate with my friends and family 4,500 miles away just as easily and cheaply as I can with friends in the United States.

Technology has also changed my contact with home in another way. It used to be almost impossible to get news from Germany. Even major U.S. newspapers have only limited coverage of foreign affairs, and the TV news even less. For many years we used to joke that it takes a major natural disaster or a revolution before one would hear anything in the news about any other country. So unless somebody told me about an event on the phone or sent me a newspaper article, I was pretty much clueless as to what happened in Germany. For many years I got by watching BBC World News on PBS (Public Broadcast Service), but then another technological change came along: internet streaming. Germany's two major state TV channels offer a significant amount of their TV programming in an archive for free internet streaming. For several years now, I have been able to watch the same news program that I used to watch when I still lived in Germany (and read the online version of the same newspaper I read at home). I can follow my favorite German TV shows, and can watch the sports events (e.g., biathlon and ski jumping) that I enjoy watching but that are almost never shown on U.S. TV. This has further reduced the distance between here and there, as I no longer feel that I live in a news vacuum and can more easily participate in conversations about German topics (I must admit, however, that this comes at the expense of not being up to date on U.S. news and TV shows, as I spend much more time watching German TV than American, and continue to prefer a good biathlon race over the Super Bowl).

ACADEMIC WORK

The technological changes also had a major impact on my academic work. In the first ten years or so of my time abroad, the U.S. was the focus of my life. I traveled to Germany for three or four weeks every year for a vacation, but virtually all my work was done in the United States. This has changed substantially in the last few years. My life is now divided between my "two homes," as I pack my bags as soon as the semester is over and head to Germany and return just in time for the next semester. Early on, this meant carrying piles of books and photocopied articles and worrying about whether I forgot anything I needed to write an article or carry out some fieldwork. Now that so much is available online, almost all articles are available in pdf format, and I can carry books on my Kindle, I can take the same bag for work-related stuff for a two-month-long trip to Germany as I do when I am physically present at the university every day.

Since I am at a teaching-focused university where there is little time for research-related work during the semester, I do a significant proportion of my research and writing when I am in Germany (including this book chapter). I had carried out a few projects (e.g., about Berlin's failed bid for the 2000 Summer Olympic Games or historic preservation in the small city of Quedlinburg) that required fieldwork in Germany in the past, but as I now spend several months in Germany every year, more and more gets written while I am in Germany regardless of where the data collection took place. My parents still cannot believe that I was able to work on a book with a colleague in Oshkosh while being in Berlin—we just sent the entire manuscript back and forth via email every couple of days and talked via Skype when there was something we had to discuss. This kind of long-distance collaborative work would have been quite difficult just a few years ago.

Despite the long periods of time I now spend in Germany every year and the amount of academic work I do there, my work is transnational in content, but not truly transnational in terms of research collaborations. My research collaborations are all with people based in the United States, so while the work may be done on different continents, it does not link people who work at universities in different countries. I would very much like to carry out projects that are not just about people who are transnational, which is itself transnational in terms of research method and research partners, but this is difficult to do since I have hardly any academic connections in Germany.

I have often wondered why I have so few links with academics in Germany. I think there are two explanations. First, when I was a student in Germany, I attended a very large university where we had almost no direct contact with our professors, so when I left for the United States, I did not have any networks with people at German universities beyond the friends I made among fellow students. Second, since I got my PhD in the United States, I have developed strong networks there. In some ways, it feels like I went to the United States too early in my academic career, before I could establish any significant connections with German academics. As a result, my "academic identity" and my academic networks are almost completely American.

In the academic literature it is often argued that a stay abroad generates transnational capital, defined as acquiring language skills, gaining knowledge of institutions, building networks, etc. However, it is less frequently mentioned that some capital gets lost—namely the knowledge of institutions and networks in the home country. This is clearly the case for me. I often feel completely out of touch with German academia. Since I left Germany, the German academic system has undergone some major reforms. To list just a few examples, the so-called Bologna Process aimed at harmonizing education systems throughout the European Union has created a credit-system that did not exist when I was a student in Berlin, American-style Bachelor's and Master's degrees have replaced the traditional German Diplom or Staatsexamen (a master's degree for future teachers), and the so-called Excellence Initiative has tried to improve the global ranking of German universities by selectively supporting specific institutions. It is somewhat ironic that in some ways the German university system has become more like the American, yet I know very little about how it works now.

Not really being up to date with the German university system is a big obstacle to a potential return to Germany. For many years I have occasionally tried to return to Germany, but it seems almost impossible to get a job at a German university. Germany has fewer universities than the United States, and therefore also fewer job opportunities for professors. Furthermore, Germany does not have a tenure system, so it is very difficult to find junior academic positions that promise job security and opportunities for advancement. It also appears that hiring decisions are often made based on factors other than just the qualifications of the applicant. As a result, many German PhDs struggle with finding academic employment. In addition to these factors that affect all job candidates, there seems to be a reluctance to hire people from abroad, and little openness towards people whose academic trajectory may be somewhat different from those that had all their academic training in Germany.

Even though I am German, it feels as if I am a foreigner when I apply for a position in Germany. I have had a couple of interviews for academic jobs in Germany, but have always felt that I was at a significant disadvantage compared to applicants who do not have foreign degrees or spent a significant amount of time abroad. On the most basic level, all my research has been published in English, and all my conference presentations were delivered in that language, so giving a research presentation in German is actually surprisingly difficult for me. It sometimes seems that in German my research sounds less impressive, as I am not able to describe it as eloquently as I can in English. This may sound surprising to some people, but since much of my advanced academic training was in English and I hardly ever read academic texts in German or discuss my work in German, I have way more practice with academic English than with academic German. While I can easily switch back and forth between thinking in English and German in daily life, it is difficult for me to speak in German about a topic that I usually think about in English. It almost feels as if I am translating in these cases, making the German less fluent and eloquent than the English.

In addition to these language issues, the demands placed on professors are quite different, so some qualifications or accomplishments that are considered impressive in the United States count for hardly anything in Germany, and I do not have some experiences or skills that are demanded of a professor in Germany. To put this in academic language: While I have accumulated substantial academic capital in the United States, some of it does not transfer to Germany or is not valued highly, and I am lacking (or have lost) some forms of academic capital that are necessary to land an academic job in Germany.

In the United States, however, being from abroad can be an advantage in the academic setting. On a basic level, I am fluent in a language other than English, and thanks to having gone through the rigorous German education system, can also read French and Spanish without too much effort. This allows me to read and incorporate the information from publications in these languages as well, something which relatively few American academics can do. Furthermore, my background is an advantage in carrying out fieldwork. For example, several years ago I joined two experts on airports in a study about the restructuring of the airport system in Berlin that I could not have done with so much ease if German was not my native language and if I did not have some knowledge about how things work in Germany. Additionally, having grown up in another country, and continuing to be closely connected to this country, sometimes provides me with topics that are interesting to a U.S./international audience simply because these topics are unfamiliar or unusual there. This is also true for teaching, as I explain next.

TEACHING

Early on in my teaching career I realized that being from abroad presents some challenges in teaching, but also offers opportunities, especially in a discipline like geography. Because I had been a student in the U.S. education system both as an undergraduate at Western Michigan University and a graduate at the University of Minnesota, worked as a teaching assistant, and audited a few undergraduate courses while finishing my

PhD, my adjustment to teaching in the U.S. was easy—I knew how a syllabus was set up, what was demanded in terms of readings and exams, and how the U.S. grading system works.

While I was well aware of how many things work in U.S. classrooms even before I started teaching and like the more student-centered approach, I must admit that I continue to struggle with some things even though they should no longer be surprising to me after teaching more than a dozen years in the United States. At the most basic level, I still cannot quite understand that many U.S. universities, like the regional university I teach at, admit so many students that in my mind are not prepared for university. In Germany, students get sorted into three different school tracks early on in their educational careers, with less than half of the students attending the *Gymnasium*, a college-prep school. Even of those, not all go on to the university, either because they choose another career that does not require a university education or because they cannot meet the university entrance requirements. Once at the university, students do not take general education classes, but immediately start working on their majors and minors. Furthermore, a typical geography university student has probably had seven or eight years of geography education at school (beginning in elementary school), and likely even majored in geography in the last two years of high school. As a consequence, the average German geography student is much better prepared for university-level geography classes both in terms of general academic skills and geography background. By contrast, many American students have not encountered geography in secondary school, and are therefore unlikely to choose geography as a major when they begin their university careers.

I still cannot help being frustrated with a good number of my students who come in with so little background and appear to make so little effort to do well in class. I continue to believe that students have to work for their education, and do not provide my students with study guides, spoon-feed them, or give extra credit, which many of them seem to expect. Even though there are of course many U.S.-born instructors who share my opinions and do not believe in making things easy for students, as a foreign instructor I often hear from students that I continue to apply European standards and do not adjust enough to what my students want. It is very frustrating to me that students sometimes blame their lack of performance on me being from abroad. Another frustration for me is that while the vast majority of students seem to appreciate hearing another perspective on world events, some students take offense. I have been told on several occasions that as a foreigner I do not have the right to criticize anything American or to voice any opinion that is not in line with what many Americans think or believe in. Like many instructors I am careful about presenting balanced opinions about issues and letting students form their own opinions, so those kinds of comments, even though rare, are really hurtful.

Despite these frustrations, and the occasional involuntary comic relief I provide my students with funny language mistakes, overall I believe that being from abroad offers lots of opportunities for making geography classes interesting. Students have told me on numerous occasions that having an instructor who has grown up abroad adds authenticity to some of the material as I can tell them details that few U.S.-born instructors would know. I am rather known on campus for telling stories in class, "stories that at first seem completely unconnected to the material but somehow end up being relevant to class," as one of my students once put it on a student evaluation form. Whether these stories are personal anecdotes of having a hedgehog in the basement over winter or what happened in political science class in 12th grade, or stories of real events such as my experiences growing up just a few yards from the Berlin Wall, they seem to interest the students and contribute to their learning. The same is true for leading study abroad trips.

Since I grew up in divided Berlin, teaching students there is always special. It is special for me, as I get to tell my students about the hardships Berliners endured during the blockade or when the Wall was there and the joy when the Berlin Wall finally fell, topics that are close to my heart. Furthermore, I can add personal stories of my and my family's experiences. Learning about these events at the locations where they happened is also special for the students, as they rarely have the chance to be in the place where history was written and hear from someone who lived through some of these events. Thus, my background and identity are important resources in teaching.

IDENTITY

For many people living outside of their home country, the question of identity is of great significance. As we know from research about people's identities, how one identifies oneself depends not only on the individual, but also on the context. For example, when somebody in Europe asks me where I am from, I will most likely respond "from Germany" or even "from Berlin," while I would probably respond to the same question with "from Europe" when asked in the United States. Likewise, depending on the context of the conversation, I may prioritize my national origin, the fact that I live abroad, my profession, or any other facet of my identity.

As I grew up in West Berlin, defining my identity was always a bit difficult. According to my passport, I was of course German (West German, or a citizen of the Federal Republic of Germany to be exact), but for many of us who lived in West Berlin during the Cold War, there were three German states—West Germany, East Germany, and West Berlin—so in our minds our citizenship was "West Berliner." Since we were separated from West Germany by East Germany, West Germany was a foreign country to us just like the Netherlands or Switzerland, except that the same language was spoken there. While I certainly cheered for West German sports teams, and considered the West German flag "my" flag, I did not fully see West Germany as "my country."

Even though I left for the United States eight years after German reunification, long after West Berliners had officially become "real Germans" and no longer lived on an island surrounded by "the East" in all directions, my identity as a Berliner persisted to some degree. While living in divided Berlin, I had probably never fully realized how unusual our political status was, but in the United States I often had to explain it to others and covered the topic in many of my classes, so I continue to strongly identify with my home city. For my students, however, I am clearly "the German," and I have come to accept that label as my primary identity.

In the migration literature, people often talk about "hyphenated identities," such as "Cuban-American" or "Romanian-American." To me this idea of being a "modified American" (where the original identity still shines through, but the identification with America is strong) makes a lot of sense, but I have never felt that I am "hyphenated." One of the reasons may be that I am still not sure whether I would like to stay in the United States permanently, so it is difficult for me to see myself as an American, even a hyphenated one. People who consider themselves immigrants and are sure that they would like to stay in the United States for the rest of their lives probably find it easier to adopt that label, as they have made a stronger break with their home country. Furthermore, even though I am in most ways well adapted to the United States, I do not feel American or even German-American. I am acculturated but not fully assimilated: I am fluent in English (in fact, I probably think more often in English than in German and write almost exclusively in English), have a good knowledge of U.S. culture, feel at home in the U.S. university system, and love the country, yet in many ways I still think, feel, and act like a German. In my mind, I am a *German in America* when I am in the United States, and a *German living in America* (or maybe an Americanized German?) when I am in Germany.

When I am in Germany I am often reminded that I am no longer 100 percent German. In many ways, I "take off my American hat" the minute the plane from the United States touches down on German soil and switch to German behaviors (such as not starting to eat before everyone has said *Guten Appetit*, or addressing people more formally) without even thinking much about it. However, sometimes I find myself in situations where I am no longer entirely sure what the "proper German way" is because things have also changed in Germany since I left. For example,

German has a formal and an informal "you" with pretty strict rules as to when each is used, but it seems to me that Germans have become more informal over time as I now hear the informal "you" much more frequently. Every once in a while I even find myself in a situation where I am utterly clueless about how things are done—for example, I once stood in a bank without the faintest idea how to do a money transfer on a bank machine because I only knew how to do the transfer on the paper form that was used when I still lived in Germany.

Despite these occasional situations where I feel like a foreigner in my home country, I see myself as German despite having spent more than a third of my life (and most of my adult life) in the United States. The question of where "home" is, however, is much more complicated for me. I actually say that I am going home regardless of which direction I travel. Berlin is home, but so is Oshkosh. They are different homes, however. Berlin, for me, is strongly associated with my family, long-term friends, and traditions, while Oshkosh is my home for work and associated with another set of friends and activities. Each trip is therefore accompanied by mixed feelings: I leave one set of friends behind, but I look forward to spending time with another set of friends. I miss certain things from one country, but enjoy many in the other. I can choose the best of both worlds, but I always miss something. I have two homes, but I am never fully at home in either place. Living in two worlds is sometimes wonderful, and sometimes confusing.

Many people are aware of me being part of and at the same time torn between places. My friends and even sometimes my students make fun of me for saying sentences like, "I will go home for Christmas, and return home three weeks later." To me this makes perfect sense. According to my students, I also use terms such as "here" and "we" in such a way. For example, in the Europe class I might say "we do this . . ." one minute and refer to "we, Germans," while in the next sentence "we" might mean "we, here in the United States." While this may sometimes confuse my students, it is actually a wonderful way to show to them how complex the concept of identity is and how difficult it can be to leave one's home country behind.

STAYING OR LEAVING?

Like many people who live outside their home country, I often think about whether I should stay in the United States, return to Germany, or even move to a third country. As described earlier in this chapter, my original plan was to get my PhD in the United States and then return to Germany to teach. At the time, I had not realized how difficult such a return would be. Looking for an academic job in the United States seemed easier, and since two of my closest friends in the United States graduated at the same time, we were able to give each other professional and emotional support throughout the process.

Like many people who graduated at that time, I was not successful in finding a tenure track position right away, but instead found myself in a one-year lecturer position at the University of Wisconsin Oshkosh. The heavy teaching load left me little time for anything but class preparation and grading, so I only applied for a few tenure track jobs elsewhere. Then a tenure track job became available at my university. I applied and was offered the job. This was probably the first time I realized that my stay in the United States would be much longer than I had originally anticipated.

In terms of personality, I am the kind of person who fully commits to something, or does not do it at all. For my professional life that meant that I did everything I could to succeed in the position I was in—delivering good classes, publishing more than required, and contributing through service. Since I still was not quite happy living in a small city (at least from my perspective—I had lived all my life in major cities such as Berlin, Vancouver, Paris, and Minneapolis/St. Paul)—I occasionally applied to other jobs, but my main focus was on doing well in my position. Time flew, and before I knew it, I was promoted to Associate Professor and received tenure one year later.

The job security offered by having tenure, coupled with the fact that I know the American university system better by now than the German, led me to believe that as long as I prioritize my professional life, remaining in the United States would make more sense. However, I was still not sure whether I wanted to stay there for the rest of my life. Because I continued to be torn, I applied for several academic jobs in Europe, and was invited to job interviews at three different universities, including one elite institution. This proved that I was competitive on the European academic job market on paper, but the interviews made it clear that my skills and experiences did not always match the demands of an academic job in Europe. Either I felt that I was lacking "local academic capital," as I did not know certain things about the university system that a native candidate would have known, or I discovered things during the interview process that made me question whether I would actually be happy in the job due to the different demands that would be placed on me. In the end, I was actually relieved that I never received a job offer from a European university as the decision of whether to accept it or not would have been agonizing.

Several years after sending the last application to a university across the Atlantic, the question of whether I will stay in the United States or return to Germany is still open for me. For the last couple of years I have prioritized my professional life, so staying in a position where I am professionally comfortable and successful seems like a logical choice. However, I may not want to prioritize my professional life over my personal life forever, my family may need me to return home, and I may not always be able or willing to split my life between two continents. Priorities change as we get older, and some topics that did not matter much in the past become more important over time. For example, I am scared by the health care system in the United States and am much more concerned about retirement in the United States than I would be in Europe. As long as one is young and healthy, both of these are minor concerns, but they will grow in importance over time. In regard to Germany, I wonder how I can take care of my parents when they need me or what will happen to the house that was built by my grandfather when my parents will no longer be able to look after it. I also fear that returning will be harder every year, so the longer I put off the decision, the more difficult it may be to realize it. So rather than the question of whether to stay or return getting resolved, it seems to get more and more complex and therefore ever more agonizing.

FINAL THOUGHTS

Since I now divide my time between living in the United States and in Germany, I can pick the best from both worlds. Furthermore, modern communication technology has made it possible to always be in touch with "the other world" to a degree that would have been unthinkable just a few years earlier. Living a transnational life can be wonderful because it offers so many opportunities, experiences, and insights that people who do not leave their home countries will never have.

However, like many people who have spent many years abroad, I struggle with the question of where I belong and where I would like to be for the rest of my life. It seems that whenever I thought that I knew what I wanted to do, circumstances (professional or personal) changed in ways that made me reconsider. My research about migration intentions made me realize that many other people are in the same situation. Even when we enjoy living transnational lives, travel back and forth regularly, and participate in both societies, ultimately the question of where we belong never goes away.

NOTES

1. I would like to thank Helen D. Hazen and Rebecca B. Theobald for their feedback on earlier drafts of this book chapter as well as for being wonderful friends and collaborators on several projects dealing with academic migrants.

THREE

Traversing Boundaries of Education and Geography

Sketching Stages, Spaces, and Standpoints of Transnational Self

Ibipo Johnston-Anumonwo

Growing up in Nigeria, before the era of Facebook, YouTube, or even the World Wide Web (I do not even recall handling a computer prior to 1981), imagine my consternation when in 1983, during my second year as a geography doctoral student, I was assigned to be the teaching assistant (TA) for Intermediate Statistics (Geography Department, Clark University, Massachusetts). I sobbed publicly for fear of failure. Ironically, riding on the crest of the patient, dedicated, and solid grounding with Perry and Susan Hanson, I ended up utilizing quantitative methods for my dissertation and for much of my early to mid-career research on gender and ethnic dimensions of locational access in cities of the United States.

I have resided in the United States since 1981. It is my home country. Yet, as I plan a trip to Nigeria, I tell my neighbors and colleagues that I am traveling "home": it is the place where I was born. I find myself spending a few moments of diversion in a new way—reflection on my past and self-assessment of my identity, instead of the typical preoccupation with the present-day tasks and obligations or attempts to plan for future undertakings. The questions "Where are you from?" and "What brought you here?" are often asked lightly, as conversational platitudes meant to elicit laconic and trite responses. A sentence about my "hometown," perhaps another about how "the mission, institutional climate and opportunities of the college really resonated with me," is all that is expected in response. But simple responses to these questions cannot fully address the complex interconnections between identity and place or indeed between multiple identities and places. The identities and lives of individuals are rooted in particular places, so like every person, the story of who I am and where I come from is crucial to understanding how I got here (and conceivably the nature and direction of my future undertakings).

NIGERIA IN BRIEF

Nigeria is defined and empowered by the very things that often cause its turbulence. Named for the Niger River, the country is the most populous in Africa and the powerhouse of West Africa. It is endowed with natural resources (including oil and natural gas), and the ingenuity of the people is well documented. Despite these facts, for many years social instability has been the norm rather than the exception. The country has suffered much division due to ethnic and religious conflict, and government corruption has stunted much of the potential socioeconomic expectations.

Drawing on African scholar Ali Mazrui's triple heritage schema, Nigeria in antiquity comprised kingdoms that would qualify as the legacy of the indigenous African heritage. Diffusion of Islam into the region from across the Sahara qualifies the Islamic heritage. The third heritage comprises forms of European influences and is partially manifested in subsequent periods when parts of coastal Nigeria became active as the "slave coast," and when European colonization took hold in Nigeria during the nineteenth century when the British advanced from the coast northwards to the interior. In the modern era, Nigeria faced many of the standard trials and tribulations of colonial rule before eventually moving towards self-governance in the post–World War II period. Even after independence in 1960, Nigeria has been embroiled in conflicts ranging from a civil war to military coups and violently disputed elections. A majority Muslim North and majority Christian South have struggled for control of land and governmental power. In recent years, non-state actors have become an increasing threat with groups like Boko Haram in the North and MEND (Movement for the Emancipation of the Niger Delta) in the South fighting for reasons both economic and cultural. However, the conflict in Nigeria is just a part of the larger story of the nation-state. When the 2015 democratic elections were once again held, the transition of power from the incumbent to the opposition party was historic for being comparatively painless. A dynamic interplay between the varied cultures, the rich histories of the various peoples, and the immense potential of the economy interweave to create a complex and unique tapestry, providing the foundation and backdrop for my formative years.

1960S–1970, THE FORMATIVE YEARS: EVOLVING NATIONAL-INTERNATIONAL CONSCIOUSNESS

In stark contrast to my lack of familiarity with modern mid/late- twentieth century information and communication technology, there was never a time in my life growing up in urbanized settings in Nigeria of the 1960s and 1970s that my family lived without a television, telephone, or car (usually two cars minimum). My parents were never unemployed. I always lived in a single-family home, often with live-in relatives; and there were years when because of their government jobs, my parents lived in separate towns and the family had homes in the two locations where my parents were employed in their capacities as civil servants. My father was a government administrator in Ibadan and my mother was principal of the Government Women's Training Center in Abeokuta. As a dual-career couple, they could afford to pay for a nanny, driver, and house helpers.

I often use my personal background for teaching concepts and topics in my geography classes. To debunk the oversimplification of Africa, I present my country of origin, Nigeria, to students as a vast, diverse, massive size a country of contrasts including rural versus urban contrasts, and a microcosm of a continent which does not have a "single story." Transnational scholar Ali Mazrui's perspective of the triple heritage fits aptly in the Nigeria where I grew up with widespread presence of traditional, Christian, and Islamic cultural elements. Nigeria is a country characterized with tremendous ethnic diversity, exemplified in its over two hundred fifty languages. I grew up bilingual, born into a Yoruba family and spoke both Yoruba and English at home.

My parents had expatriate colleagues and international friends from England, the United States, Ghana, Jamaica, and India, to name a few. One could infer that some were in Nigeria as part of the euphoria of post-independence, and many Americans were certainly there in the spirit of international service during the beginning of the Peace Corps era. Among my mother's many international visitors in the 1960s and 1970s, were (Auntie) Gwen Cocker from England; Janaki from India; Marian Hughes and her husband, Paul Hughes, from the United States (African-American), principal of Abeokuta Girls Grammar School who returned to Albany, New York; David and Audrey Radmore (Peace Corps volunteers teaching physical education and art who returned to Stockton, California); Miss Mary Ann Zall (also Peace Corps); and researchers, for example, Ms. Hilferty (Massachusetts) and Professors Joanne Eicher (Minnesota) and Henry Drewal (Wisconsin).

I have kept in touch with them all in rare visits and mostly through occasional phone calls or emails, as lifelong family friends. Apart from Janaki, whose explicit knowledge of the reputation of development geography professor and department chair Len Berry informed my decision to apply to Clark University in 1982, there was no professional connection with any of the others in terms of my U.S. education and training. My ties with them have been in the form of deep, sustained support that has positively influenced my experiences of living in the United States with the knowledge that emotional outreach of genuine friendship of Americans is only a phone call away.

PERSONAL BACKGROUND: HYBRID(IZATION) AND TRANSNATIONAL CULTURE IN A GLOBAL AGE

Both my parents were born in Lagos, and Lagos is my hometown, although when I was born, they lived outside Lagos. The names of places, toponyms, in Lagos underscore the significance of multiple influences, internal and external. Among the neighborhoods and locales with international place names in (pre-1960s) Lagos are Popo Aguda—Catholic Quarters (where mostly Brazilian repatriates lived in the late 1800s). Along the same vein, names that are significant in my background are a combination of traditional Yoruba names and those that are not of traditional African origin, among which are Johnston, Williams, Howson-Wright, Smith; or those of other Lagosians such as Cardozo, Pereira, Cruz, or Domingo (i.e., Portuguese names)—a testimony to the trans-Atlantic slave trade; subsequent return migration, resettlement, and the resultant familial links outside of Nigeria, for example, with Freetown in Sierra Leone. Returnee ex-slaves (known as Creoles), who settled in Lagos from Sierra Leone, the West Indies, and also from Brazil (as skilled artisans), contributed to the city's modernization. Brazilian influences in Lagos are evident in the architecture, carnivals, and food (e.g., Robin Law). Many Lagos families, including mine, have "Saro" (i.e., Sierra Leonean) ancestry.

My name, Ibipo, in Yoruba means "large family circle." There are, in fact, a large number of relatives as part of my extended family, but the size of my nuclear family is small. I am one of three children in a monogamous family, which was not typical, since having four or more children in the family was common in my generation, and many Nigerian families are polygamous. Regardless of monogamous or polygamous leaning, Nigerian society is patriarchal by and large, although a strict male-dominated perspective was not evident in my extended family in general, or between my parents in particular, undoubtedly because of their above average educational background.

On a national scale, the significant event between 1967 and 1970 was the Nigerian civil war, wherein the Eastern region, Biafra, aimed to secede, took place during my elementary school years, as well as the geographic reconfiguration of the country in the form of the 1968 creation of states. At the time that I started secondary school, the civil war ended. By then, my mother had returned from a year's trip based in Albany, New

York, during August 1967 and August 1968. U.S. influences became increasingly added to British and Nigerian ones as part of my youthful identity. At home we knew Grimm's fairy tales *and* tortoise tales; games included "Ayo," Ludo, snakes and ladders, Scrabble, jigsaw, netball; much-loved TV shows ranged from *Bonanza* to the *Village Headmaster*; favorite non-Nigerian books were authored by Agatha Christie, Enid Blyton, and James Hadley Chase; and our mother's extensive book collection included the (U.S.-based) *World Book Encyclopedia* rather than *Encyclopedia Britannica.*

The family had relocated from the relatively smaller, laid-back historic cultural urban center, Abeokuta, to the national capital Lagos, which was growing both from geographical sprawl/annexation and from natural increase/in-migration, at a pace which pales compared to subsequent decades. For my secondary school education, I attended Methodist Girls High School, a single sex (versus co-ed schools) day-school (versus boarding school) that had been founded in the late 1870s as a mission school (versus non-religious affiliation schools) but had become government operated at about the time when I enrolled in 1970. Administered by the new Lagos State government, I enrolled when the last of the "expatriates," Miss Walker from England was the school principal, with only one or two other Europeans as members of staff (e.g., Miss Dentex, from France, who taught us French). Before I completed high school, the school had its first Nigerian principal, Mrs. Odulana, reflecting the new era characteristic of post-independence Nigeria.

Both at school and at home, my upbringing was quite multicultural/transcultural. Subjects and activities at school (during the 1970s) were English, mathematics, biology, literature, French, chemistry, geography (we learned about Africa too, not simply the British Isles), scripture, domestic science, and needlework (males did technical drawing and/or wood work). But typing was not one of the subjects taught in schools like mine—I still cannot type till today. In literature, we read the works of Austen *and* Achebe, Shakespeare *and* Camara Laye (African Child), Grimm's Fairy tales *and* tortoise tales.

My mother instilled in us a very healthy dose of Yoruba culture. She hired a traditional drummer to give us drum lessons. "*Iyabo of Nigeria*"—the title of a (1973) U.S.-published book written by my mother, followed by a manuscript, "Bringing up Iyabo thru Yoruba proverbs," were among the undertakings and writing projects that my mother undertook that highlighted traditional Yoruba values like respect for/learning from elders/others central in our upbringing. At family gatherings—weddings, christenings, funerals, housewarmings, birthdays, picnics, and so forth, there was an explicit mix of traditional Yoruba and British influences. Because of my relatively slow foreign language acquisition ability, my mother paid a Nigerian bilingual young mother married to a pilot as a tutor so I could pass the high school French exam.

During the 1970s, my awareness of and exposure to international concerns intensified—in the form of subjects about racism in apartheid South Africa and U.S. civil rights. Nigeria was enjoying the economic benefits of an oil boom and spent its revenue (petrodollars) for infrastructure that, in Lagos, consisted of the construction of a second mainland bridge. Nigeria hosted multilateral events like the All Africa Games (with the attendant construction of the National Stadium in Lagos), as well as the International Festival of Arts and Culture (Festac 1977) during which I saw performances by Stevie Wonder and Osibisa. I recall the screening of *Roots* at the National Theatre—an edifice inspired by Eastern European architecture. The dynamism of the 1960s and 1970s was reflected nationally and internationally.

Unlike my parents' generation when students had to sit for British (Cambridge)-based exams, my own secondary school completion exam was administered by the West African Exam Council (WAEC), and as such is a transnational school certificate covering Anglophone West Africa, Nigeria, Ghana, Sierra Leone, and Gambia. Yet features of the British system remained (i.e., "O-levels" GCE; HSC and A levels"). In secondary school and university, few tests were multiple-choice type questions; there were no "curves," and no extra credit. Exams during my university years were largely summative—at the end of the term with a "do or die" aura.

TRANSNATIONAL TRANSDISCIPLINARY TRAINING IN EDUCATION AND GEOGRAPHY

University years: 1974–1978

University of Ibadan (UI), where I earned my first degree (B.Ed.), is widely considered Nigeria's premier university and was a Nigerian Federal Government University. UI was the first university in Nigeria to adopt the United States' credit system. I attended for 4 years—1974 to 1978, the first year of which was for Preliminary Studies ("Prelim")—a competitive accelerated program that reduces the two-year A-levels to one year, and eliminates the need to sit for the HSC exam. Classes were taught by the university professors themselves, not teaching assistants (TAs). Most were men. A couple of examples are: Professor Akin Mabogunje, of international repute as past president of the International Geographical Union (IGU), and Professor Reuben K. Udo, author of a premier textbook on the geography of Africa—and professors in teacher education that contributed to the robust preparation and solid credentials that sealed my future admissions to Harvard and Clark Universities. A retired senior colleague, the late John Wilmer, observed in 2012 the significance of an ingredient of my academic training—"being versed in the knowledge of first degree content of the study of academic 'British Geography' taught by staff members at the University of Ibadan who had been trained at the University of Nigeria, located in Ibo land, pre the civil war!"

John Wilmer, who played a decisive role in my academic career as a geography professor here at SUNY Cortland, had taught at the University of Nigeria from 1961 to 1966. Unlike the professors that taught me, (like Pius Sada—a former honors student of John Willmer), who had many European professors, I had only one European professor during my 4 years at UI, namely Father Foley, for my prelim chemistry class in 1974/1975. But the majority of my professors had received their geography training abroad, especially in the UK and increasingly in the United States. For example, one of my education professors was associated with the U.S. High School Geography Project (HSGP) of the 1960s/1970s and included many components of this curricular initiative in course lectures; while another, Dr. Stanley Okafor, was the one who (when in 1982 I went to tell him about my graduate admission to Clark University) showed me a copy of the (green) book, *Radical Geography* by Dick Peet, from his office bookshelf. A transnational spirit was evident in UI's academic resources, personnel, and content.

OVERSEAS VACATION IN ENGLAND

I had a chance to travel outside Nigeria for the first time in my life in 1976 when my mother paid for a vacation trip to England. I spent about a month visiting with my uncle and his close family friends who had lived in London since the 1960s. I also spent some of my vacation with another female relative of my mother's also in London; with my mother's former teacher and close friend, Auntie Gwen, in Huddersfield; and much time with a few Nigerian friends and university mates who were also vacationing in London during that summer. (In the 1970s, spending part of the long vacation from school abroad, especially in London, was steadily becoming a trend among middle class Nigerians.)

I had read many of the murder mystery novels of Agatha Christie and wanted to see one of her widely acclaimed plays rather than the one my aunt suggested. Reluctantly, I went with her and saw her selection—the South African production of *Ipi Tombi* instead, much to my great delight. It constitutes a subtle transgressive slant in my consciousness, in the sense that it lay open to me the conditions of repression as well as forms of resistance during the apartheid era by showcasing the excellent creativity and talent of African playwrights, musicians, and actors. I left the theater not regretting that I did not see yet another classic European production instead.

LEAVING NIGERIA: AN INDIRECT COURSE TO A U.S. GEOGRAPHY DEGREE, 1980S

Having vacationed there, traveling to England was not a novelty to me anymore when I decided to study abroad after my first degree. I applied exclusively for admission to U.S. institutions, primarily out of curiosity. I had moved back home to Lagos after completing my youth service in 1979. I lived with my mother while working as a geography teacher at Ikeja High School for two years. There were U.S.-based information resources within easy access in Lagos, such as the USIA library. In the 1970s, I regularly accompanied my mother to the annual JFK essay competition awards ceremony held at the Nigerian Institute of International Affairs (NIIA), since she was on the awards selection board; and I had developed a fondness for a dedicated library—the Whitney Young memorial library in one of the neighborhoods in mainland Lagos. That was where I studied for my GRE and TOEFL exams. Because I was applying to grad schools, I did not have to take the SAT, unlike my older brother who took the SAT years earlier while seeking admission abroad for his undergraduate degree—even though he ended up attending a college in Canada rather than in the United States.

Unlike my superior scores on the TOEFL test, I attempted the GRE twice and managed to earn satisfactory scores. In one of the attempts, for the subject (discipline) portion, I took the geography test and in the other I took the education test. I planned to apply exclusively to geography programs and chose about a half dozen institutions including Johns Hopkins and Madison; but when a cousin suggested that I should consider applying to Harvard, I followed his advice, even though I told him that I thought it was an unrealistic consideration as an Ivy league institution and that the competition for gaining admission at Harvard would be far too intense. In any case, Harvard had no geography program, so I applied to the education school instead. Of the half dozen graduate schools to which I applied, the only one that I was accepted by was the Harvard Graduate School of Education (HGSE). All the geography programs turned me down.

With a letter of admission from the HGSE, I applied for and was able to get a Nigerian federal government ministry of education scholarship for the one-year master's program. I was given a three-year study leave from my teaching position where I had taught high school geography between 1979 and 1981. And then with my student visa (J1/F1) in hand, I left the position in August 1981 for the United States to study for a master's degree in education. I did not return to the high school teaching position, thus constituting part of the statistics considered the African "brain drain."

AFRICAN IN AMERICA: COPING WITH RACE, NATIONALITY, AND "FOREIGN STUDENT" IDENTITY

The trip and arrival in the United States was not particularly unusual. It was the lingo and vocabulary about race, skin color, and ethnicity that puzzled and confused me most. I remember saying something along the lines of, "If they call someone with such a light complexion black, then what would they call a dark-skinned person like me, charcoal?"

Living and going to school in Cambridge, Massachusetts, next to pluralistic and diverse Boston, offered huge opportunities to be with international students. My off-campus roommate during the year was from Lesotho. At college, I met for the first time somebody, Lydia King, from Puerto Rico (Spanish-speaking, "Hispanic," was a new term to me then) that led to a friendship that has lasted a lifetime. Adjusting to life in Boston was not difficult, especially because there were many Nigerians there including a family friend who met me at the airport and had made arrangements for my off-campus housing. But a key challenge was how to negotiate classroom encounters both in terms of my accent and the American manner of speech. Nor was I used to the teaching method of discussions and speaking up in class. In the first few days during orientation and other meetings, I had the sinking feeling that I was never going to understand or catch up with the speed at which Americans speak. But that huge sense of insecurity faded within a few weeks.

A significant part of my adjustment is attributed to a Nigerian classmate, Nneka Dike, whose father was a retired professor at Harvard, and she herself had attended both high school and college in Boston (her family belonged to the Biafran diaspora of the Nigerian civil war in the 1960s). Her friendship allowed a very smooth slide into the pool for Lydia and me—both of whom like Nneka are "black." At the Ed. School, we cultivated good support friends from Liberia and Kenya as well as among other US students—white and other ethnic groups and from other non-U.S. countries in our eight courses, some of which were about educational planning in developing countries and education research in the third world. Nneka and others saw us through our first snow, research papers, menu selections at restaurants, and much more. I finished the fall semester with two A minuses and two "A's." I was pleased and relaxed, and better able to enjoy my courses in the spring semester with reduced anxiety.

One of the lovely sunny days in early spring gave me an opportunity to take a leisurely bus ride from Cambridge without really knowing where the bus was heading. It was an eye-opening experience for me. I had been in Boston for about eight months since August. I realized from the bus trip that at every stop down the route more white people got off and more black people came on. Prior to that day in April, I had not seen this many black people in one particular location or at any particular time. I stayed in the bus all the way to the terminal in Roxbury and watched the reverse pattern that took place on the return trip back to Cambridge as many black people got off the bus and many white people came on. Many years after, it still is mind-boggling to me how I could have conceivably returned home to Nigeria with the possibility of saying that I hardly witnessed any residential segregation or blatant forms of racism after I had lived in the United States for eight months, precisely because these patterns are somewhat "hidden" and not immediately visible or apparent to the average person who resides in a predominantly white or predominantly black section of a city. Living together apart, indeed.

I question to what degree can a (foreign-born) black person really feel that she/he belongs in the United States? My experiences and encounters of overt racism are indeed *very* few and far between, but instances of prejudice and discrimination based on an identity as a foreign-born black person mean that bothersome questions about my transnational status and identity remain in my mind.

CLARK GRAD SCHOOL 1982–1987: FROM WAITING LIST TO GEOGRAPHY AND GENDER DISSERTATION

When I applied to Clark University Graduate School of Geography in Worcester, Massachusetts, for my PhD, I was not admitted at first. Instead I was on the waiting list, and afterwards I received the offer to enroll in the fall. I returned home to Nigeria during the summer, and I also processed my student visa papers for the PhD studies at Clark University. It was while I was home in Nigeria at my alma mater in Ibadan that one of my former U.S.-trained professors (mentioned earlier) gave me one of my memorable introductions to Clark's geography international renown.

Another early recollection that I have is the agreeable rapport between (American) U.S.-born and international (foreign) graduate students. One American student, Dan Weiner, paid for the housing deposit on my behalf without ever meeting me face to face. I was very grateful, especially because Clark University did not make any institutional provisions for international students' housing. Along with my classmate Ann Oberhauser, the friendships among Dan, Ann, and me went past housemates to lasting relationships as family friends. Ann, and later Dan's brother, visited my family in Lagos during trips to Africa, and I have been at their several family homes and many events. In 1984 while still in graduate school I traveled home to Nigeria with my fiancé for our own wedding. We were both graduate students at the time so a short trip to Nigeria in December and back on our student visas posed no problem and was convenient.

I often recall the graduate school years for fostering a healthy working relationship and an equally cordial out-of-class relationship among and between graduate students from many different countries. Unlike my undergraduate years in Nigeria, when there were very few international students enrolled at the University of Ibadan when I attended in the 1970s, I recall that grad school in the United States was where I met the most diverse set of international students in my life. My entering class in 1982 at Clark with twelve students was supposedly one of the larger cohorts. In addition to including four females, there was also representation from Nigeria, Tanzania, Morocco, and England, to name a few countries. Indeed it was in grad school in the United States that I first met individuals from Swaziland, Spain, Zambia, Zimbabwe, Cameroon, Israel, Costa Rica, and Japan. Coupled with students in the master's degree program for international development, there were more from Iran, Turkey, Kenya, Uganda, and South Africa. Such a vibrant presence of international citizens was a major aspect of my grad school experience. Some of us graduate students got RA assignments in the Center for International Development (CENTED). One of my assignments was to write a country profile of Somalia. Research assignments such as these, and even more important the collaboration with international students, elevated my global consciousness and awareness. I was told (years later by a senior graduate student who had served as the student representative on the admissions committee during my entering year) that although my first and second degrees in education rather than in geography had contributed to my being wait-listed in 1981, it was indeed my extensive geography coursework and academic background from schooling in Nigeria that also kept me in the pool. After two degrees in education, I was now at Clark University on course for a PhD in geography.

I had to cancel my incipient plan to do my doctoral research on urban Africa via fieldwork abroad. I wanted to work with Dr. Susan Hanson, but at the time she was not doing any research in African cities. Instead, I capitalized on her expertise; I changed to a topic on U.S. urban cities and gender, using secondary data. As a pioneer in feminist geography, Dr. Hanson guided my empirical research on geographic perspectives on gender differences in urban employment access. The exposure to taking and being a TA in Perry Hanson's statistics classes was invaluable for analyzing census data during my dissertation research and afterwards. A highly accomplished geographer and past president of the AAG, Susan remains an excellent mentor to me. I was Susan Hanson's first doctoral student. My committee was conceivably the first all-female doctoral dissertation committee in U.S. Geography, yet prior to now, I did not consider the transnational sensibilities or composition of the three women members of my committee. A closer look follows.

Susan and Perry Hanson carried out pioneering research on gendered journey to work behavior in Europe, specifically Sweden. They had lived for two years in East Africa and had their first child in Kenya while serving as Peace Corps volunteers in the 1960s; and they also welcomed me as a household resident during my Clark days in their Worcester home. Originally from Canada, Gerry Pratt arrived (and stayed) at Clark at a time when I could work with her on my dissertation, along with my advisor, Susan Hanson, before she returned to British Columbia, Canada. She is currently a leading internationally renowned geographer conducting research in transnational feminism. Briavel Holcomb, in Rutgers University, my external examiner, also is a transnational geographer originally from the United Kingdom. It is inspiring that now I am following in her footsteps as an AAG distinguished teaching honors recipient. When I completed my doctorate in 1987, they all remained members of my diverse network of largely transnational geographers and non-geographers on whom I can depend professionally. Through the university, I was able to apply for a renewal of my student visa, and the extension was granted. In effect, this transition period ensured that at no time during my stay in the United States was I without a valid visa. My consistent legitimate visa status served me well fortuitously a year later when I opted to visit home in Nigeria in 1989 for a chance for my mother and other family members to meet my daughter (one and a half years old then) after I had completed one year employed as a college faculty member.

WAITING FOR TENURE: FROM FOREIGN JOB APPLICANT TO TENURED FOREIGN FACULTY

In the spring of 1988 I was still looking for a job. I was living in Syracuse, New York, with my husband and newborn daughter. The AAG conference that year was being held across the country in Phoenix, Arizona. I was in no position to travel to what would normally be a focal point for networking, education, and professional development. A fellow student of mine at Clark, who is also a very good friend of mine, called me on the phone to tell me about a potential position at the Geography Department of SUNY Cortland, in Upstate New York, barely half an hour south of Syracuse. She, unlike me who did not have the financial means at the time, had a subscription to the AAG newsletter; she mailed me a clipping from the JIG (Jobs in Geography) section of the advertised job. Keenly conscious of linguistic profiling as a possibility in hiring, I vaguely recall what might be considered a "subversive"/transgressive ploy of a friend with a U.S.-native accent (i.e., non-foreign) calling on my behalf, as Dr. Ibipo Johnston, very deliberately leaving out the Anumonwo in my married (hyphenated) last name, to find out if applications were still being accepted; and when the answer was "yes," I put together my application materials for submission. A few years later, I was told that department staff members wondered whether, with a name like Johnston, I was a Swede!

I drove 40 minutes south on I-81 to deliver my application letter and CV in person to the geography department. Three of the five (all male) faculty members were present. They were the senior faculty members. I mentioned that I was Nigerian. Immediately, this caught the attention of Distinguished Professor John Wilmer. Wilmer, born in England and who previously lived with his wife and four children in West Africa (he had taught in Ghana and Nigeria during the 1950s and 1960s), was eager to engage with me. I was in disbelief that in such a small town, thousands of miles from home, I was talking to a man with an understanding and appreciation for my academic preparation and cultural background.

Following declined offers or disappointing performances by the three interviewed candidates, I was invited for a campus interview given my proximity to the campus. I got the job. Looking back, I should have negotiated for a higher salary. Maybe I could have received more benefits. Maybe I was too eager to jump at a job in my field, close to home, and with familiar faces nearby. Perhaps as a non-U.S. citizen, or not being brought up here, or not being assertive or "because women do not ask," I simply accepted the (comparatively low) salary that I was offered. I have stayed at Cortland for 27 (almost 30) years now. Although I taught abroad at the University of Lagos during my sabbatical leave in 1996, and I have taught as an adjunct faculty at two colleges in Upstate New York—SUNY, Cortland is my home institution. And through the ups-and-downs expected from any career, I can't say that I could have picked a better institution. Collegiality and caring surpass cash compensation. A hospitable *esprit de corps* prevails in the geography department, and I have been a beneficiary many times, such as colleagues teaching my classes during the premature birth of my son or showing genuine concern and offering practical assistance when I needed support traversing personal and professional hurdles as an African/American woman.

The process of becoming a green cardholder, (i.e., resident alien) takes an emotional and financial toll on the foreign-born academic when there is inadequate institutional expertise, knowledge or financial support. The time crunch and the legal fees are significant burdens. Amidst the tenure-clock vulnerability, I was able to develop an integrated professional portfolio with seamless connections across geography, international studies, and multicultural and gender studies in teaching, research, and university, professional, and community service as a platform upon which I earned promotion and tenure as I navigated the official transition from highly skilled H1 visa to resident alien status (i.e., green card). Illustrative activities as part of my work at SUNY Cortland are serving as a member of the International Studies Faculty Advisory, Pan Africa Project, and supporting student groups like Pan African Students Association, Women of Color, and International Awareness Club.

TRANSNATIONAL STANDPOINTS IN TEACHING, SERVICE, AND SCHOLARSHIP

To illustrate transnational professionalism. I made some choices because of, or in spite of being a foreign-born individual with transdisciplinary transnational training in education and geography. A few examples of choices with transnational bearing that I made in my teaching, service, and research work as an academic geographer follows.

In addition to upper level courses on the geographies of Africa, I teach an introductory course, "Human Geography and Global Development," a General Education course in the category for Contrasting Cultures, and I blend in pedagogy for transnational awareness and global citizenship. Increasingly, I find myself asking students: who are "we"? And I assign them to read classic pieces (Nacirema) and write essays titled "A Tale of Two Countries" as a way to shed light on comparisons with the United States and to counter ethnocentrism.

My "gender role constraint" as a mother partly contributed to an inability to pursue a career of field research (in Africa or Nigeria), which meant that I was unable to develop any primary data-based scholarship. However, in retrospect, I was able to navigate this personal/professional hurdle. I had put my research on African development on the sidelines. I do not have any research travel opportunities in Africa. But dating to a Ford Foundation–funded Summer Institute on Women and Development while I was a graduate student at Clark University, that was held at Madison, Wisconsin, I have kept a keen scholarly interest on the theme of gender and development in Africa primarily by reading and engaging in other pursuits such as editing. It was not until after I was almost tenured that I activated scholarly work in African studies when an African geographer, Samuel Aryeeytey Attoh at the University of Toledo, invited me to write a chapter on gender and development in his edited textbook on geography of sub-Saharan Africa.

I sought alternative avenues to keep active on a research agenda focused on gender and development in Africa—primarily via joining and serving in executive board capacities in a regional African Studies group; editorial work consisting of international geographers; and by sustaining an interest and service in professional women's organizations centered on gender equity and social justice concerns. As part of my professional community service roles, I applied and was selected to represent the U.S. delegation at the World Council of the YWCA in 2007. The trip allowed first-time travel to South Africa and Kenya for two weeks (accompanied by my daughter). I am also a member of Zonta International, another professional organization whose mission is improving the status of women worldwide. I have relatives and family friends whose leadership in Nigeria-based clubs of both organizations dating back to my youth undoubtedly influences my decision (as an adult professional woman in the

United States) to be a member of both of these international women's service organizations.

My recent publications on transnational feminism are edited works with a sizeable fraction of international non-U.S.-born contributors, and my current collaborative research on transnational migration studies Eritreans in Tel Aviv. Some of my professional service with a transnational dimension include editorial board work for the journal of transnational women's and gender studies—*WAGADU*, and the *African Geographical Review*.

TRANSNATIONAL TRAVERSING AS AN *AFRICAN/AMERICAN* WOMAN

I became a citizen of the United States in August 2007. On the day of my swearing-in ceremony, I spiritedly and symbolically wore a Yoruba outfit with red, white, and blue colors. I was supported by my children, and in the company of other individuals and their family members all overjoyed to be at the end of the long journey that leads to gaining U.S. citizenship. The heralded "blue passport" would be mine for the taking. I already had a passport application filled out, and I dropped it off at the post office on the way home. Still, I couldn't help but wonder just how much my life would really change. I had lived in the United States for over twenty-five years by then; what did "official" citizenship really mean?

I remember thinking back to the content of the citizenship test that what I was learning was a relevant but incomplete part of what it meant to be an American. Granted, an understanding of government and policy is important, and knowledge of the rule of law essential, but none of this struck me as "my America." I ask my children to join in reflections of transnational identity. They note that America was what I had lived for the past thirty-four years. It was partly driving my children from band practice to soccer games and picking up pizza on the way home from Girl Scouts and Boy Scouts. And yet, there was more still; America was not some compartment of my life, separate from the rest. Rather, my transnational status and upbringing means that I incorporate America into my Nigerian mindset. Trips to Syracuse were not just chances to take my children to the mall, away from almost exclusively white Cortland, but also an opportunity to head to the ethnic food store which occasionally had some Nigerian food items in stock, and now frequently does. They may eat pizza from the mall food court for lunch and then come home to joll of rice and fried plantains for dinner. I could see that this juxtaposition of the two cultures by which I had been shaped would be a recurring theme in the life of my children. What then would it mean for them to be Americans? They were born and raised in the "American" culture, to be sure, but will calling them American feel disingenuous? This is a question that only they can answer, and I believe that their own identity and perception will change with time.

"WELCOME HOME"[1]

For me, an ongoing question is: do I feel at home in Nigeria, or do I feel at home in the United States? The answer requires examining aspects of professional and personal realities that bother me or appeal to me in either location. For example, I do not miss the decayed educational facilities or delayed salaries for university professors and schoolteachers in Nigeria. I contrast the anxieties inherent in me being a mother and bringing up children in the U.S. with the relative relaxation of that gender role for my mother doing the same in Nigeria. My life as a mother caught in between cultures, in a household of two academic-career parents who are constantly on the run, from the college classroom to dance classes to soccer practices and to conference halls, is a contrast to my own more grounded mother, who had a less harried life while I was growing up as a Nigerian in Nigeria. Now that I have a U.S. passport, I have experienced some changes in the ability to navigate obstacles. Travel is as expected much easier; the freedom offered by U.S. citizenship cannot be overstated. Although the instances of delayed encounters at the airport has reduced significantly when I travel with my U.S. passport, I still feel that I am stopped and monitored more frequently than the average person at airport checkpoints. Given my Nigerian name, my bags are still "randomly selected for search" when I fly. But visa applications for international travel are now less onerous. Even domestically, applying for items such as driver's licenses or replacement birth certificates for my children is easier. And when I return from overseas travel, the immigration lines are shorter and the distrustful questions asked fewer.

In short, I may no longer have to jump through tenuous hoops to function as a person or a professional in the United States. Since gaining U.S. citizenship I have attended conferences internationally in countries such as Brazil, India, Ghana, Ecuador, and Peru with the ease that a U.S. passport enables, all the while secure in the knowledge that I have two appealing countries that I call home at the same time, modern telecommunication (i.e., Facebook, Skype) has helped with *feeling at home* and *having two homes*.

NOTES

1. Immigration Officials' typical greeting to U.S. Citizens at airports.

FOUR

Mobility, Time, and Home

The Experience of Two Generations of Transnational Chinese Geographers

Shaolu Yu and Wei Li

The co-authors of this chapter belong to two different generations of Chinese geographers born and raised in the People's Republic of China (PRC or China) but who currently live and study or work in the United States.[1] Their collective, but also divergent, academic journeys reflect the changing realities of transnational activities and connections between the two countries of their birth and current residence over the past few decades.

Wei Li grew up in the PRC when it was an isolated socialist country where anyone with a connection to a foreign country could be perceived as suspicious, or worse, as a potential spy for foreign interests. It was not until after 1978 that the door was gradually opened to large numbers of Western visitors, especially Western academics. Her transnational connections began when she became a master's student at the Geography Department of Peking University in Beijing in 1982. During the six years when she studied and later worked there, she was selected by the department, among a few other colleagues, to serve periodically as an assistant, an interpreter, or even a companion for scholars and their families during their lecturing and sightseeing trips on and off campus. Her own transnational journey began on March 1, 1988, when she was selected by Peking University to serve as a junior exchange scholar at the University of Nebraska-Lincoln (UNL) and flew across the Pacific Ocean for the first time.

When Wei came to the United States the first time, Shaolu Yu was just a young child. Born and raised in 1980s China, Shaolu had experienced the aftermath of the Cultural Revolution, the One Child Policy, and the economic reform in China in both her and her parents' life. The idea of going abroad to pursue her PhD brewed in her mind when waves of her peers in college were preparing for the GRE and TOEFL[2] in order to be admitted to graduate programs in the United States. The idea grew stronger when the transnational connections at the institutional level began to increase. Every summer, her department would invite professors from the United States to conduct summer workshops and give lectures. These professors, most of whom were born in China and settled in the United States after obtaining their PhDs, brought textbooks, research trends, and life experiences in academia in the United States. She wanted to become one of them. Her own journey to the United States began on August 13, 2010, when she left China for the first time. By then Wei was already a professor at ASU, whom Shaolu admired as a role model. Now Shaolu has earned her doctoral degree in geography from the University of Connecticut and is currently a postdoctoral fellow at Rhodes College.

While both transnationalism and globalization imply increasing spatial mobility and connections, transnationalism is distinguished from globalization by stressing heterogeneity and problematizing border, place, and identity as contested between here and there, then and now, insider and outsider (Smith, 2001; Levitt, 2001; Ma, 2003; Freidman and Schultermandl, 2011). What do border, mobility, home, place, and identity mean to the two-generation Chinese geographers in the United States? Through transnational autobiographical accounts and dialogues, we examine the similarities and differences of transnational experience over two generations, what has changed and remained unchanged, and how the transnational experiences have been inscribed in both our personal and academic lives, which reflects the academic and transnational contexts across space and time. In this chapter, through the presentation of and illustration of three different but related subjects, we document the changing mobility and immobility of Chinese-born geographers in a globalization era; the changing sense of home as transnational geographers; and the changes of a transnational place through time. We consider UConn as a transnational place given its increasing presence of international students, faculty, and collaborations. It is also the same institution that nurtured the transnational experience of both authors. The chapter will be organized via a dialogue between the co-authors, and will piece together some commonalities and distinctions.

CHANGING MOBILITY AND IMMOBILITY OF PRC-BORN GEOGRAPHERS

Even in this globalizing era with the "global race for talents" and despite the claims that globalization eradicates the importance of national borders and provides almost "footloose" mobility for highly skilled migrants (Hamilton, 1999; Ong, 1999; Bauman, 2000), transnational mobility is still an elusive dream hitting against a harsh reality. Physical mobility, or lack thereof, depends on one's visa, citizenship, and nativity status, but has little to do with one's human capital level and/or socioeconomic status, as Tsuda (2011) demonstrated among Japanese Brazilians working in Japan. The physical immobility can hinder one's social mobility and is embodied in everyday life.

Shaolu—The distance between the Netherlands and Connecticut is 3,700 miles, a seven-hour flight. But for me, it means one additional trip to Canada, two trips to New York City, and three applications for a visa which will take two to three months to process. Transnationalism and mobility are the two major themes of my doctoral dissertation, but are ironically the mirage and illusion of my personal life.

In December 2013, my paper was accepted by a conference in the Netherlands. My excitement was beyond words as this meant the recognition for my hard work, a trip to the Netherlands, a possible publication, and a great opportunity to network with scholars in my field. Unfortunately, the excitement was soon overwritten by the price I had to pay to make the trip happen; not only the monetary price, but more importantly the time and effort to fill out numerous forms and documents so that I can legally cross the border. Being an international student in the United States from China is not easy. For the Chinese students who entered the United States prior to November 12, 2014, their F-1 student visa can only be held for one year (the new F-1 visa policy extended the validity up to five years). Visa expiration means being stranded in the United States. In order to go to the Netherlands, I have to renew my U.S. visa; in order to renew my visa, I have to go to Canada (or another country as an F-1 visa cannot be renewed inside the United States); in order to go to Canada, I would need to make two trips to New York City to apply for my Canadian visa. Only after all this is done successfully can I finally apply for the Schengen visa[3] that allows me to go to the Netherlands. This was a puzzle, and I failed in my second phase because the trip to Canada had already taken most of my available time and money before the conference. I still remember the pain of having to cancel my conference trip to the Netherlands.

Two years later I was in the job market. As my passion has always been for teaching and research, I was determined to focus on academic jobs only. After five years of training as a doctoral student in the United States, I decided to stay and build my career here, continuing my trajectories of research on race, ethnicity, and immigration. However, my dream also came with the challenges of settling down as a foreigner and living the rest of my life far away from my family and my homeland where I spent my youth and early adulthood. Many times reading a job post felt like a rollercoaster ride. I was so thrilled by the job descriptions that fit me perfectly only to find them followed by a cold sentence: "U.S. citizens and permanent residents are preferred (only)"; or "Visa sponsorship is not available for this position." Ironically it was often followed by the claim that "XX is an Equal Opportunity/Affirmative Action Employer," which, to me, suggests equality in the job market regardless of color, nationality, or gender. One time, I got a phone call from a college inviting me for an interview. I had not fully digested the excitement yet before they made another call saying that they had to cancel my interview because they could not provide visa support. Only then did I realize that what limits my job opportunities is not my capability but my citizenship. Later I also learned that if I could not find a job or unpaid volunteering work associated with geography within the sixty days of grace period after graduation, I had to leave the United States. This further limits my flexibility in job searching as I have to take on anything I can find in order to maintain my legal status. The last semester of my doctoral life was filled with the excitement of finishing up my dissertation and ambitions to start my new career, but in the meantime haunted by stress and frustrations on the job market, fear for the future uncertainty, and the guilt for my parents who were both reaching their sixties, especially as the only child in my family. For the past five years during my doctoral program, I only got to see them twice, due both to the busy schedule as a doctoral student and notorious visa renewal process. Fortunately, I finally landed a postdoctoral position in the South, which fits my research and teaching background very well. But that leap in my career is not accompanied by physical mobility; rather, I got more confined with limited physical mobility due to the Optional Practical Training (OPT) status.[4]

I have heard many stories from my Chinese peers that their visas were checked, delayed, or even rejected while traveling to China with OPT. It was also suggested by my lawyer that I'd better not travel abroad at all because that would put me at risk of not being able to come back to the United States again, regardless that I have a job here. What it means to me and my family is that I will not be able go back to my hometown for four consecutive years since 2012. As a geographer, I am thrilled by the fancy concepts of the mobility paradigm, time-space compression, global village, and flexible citizenship. One of the questions I have for my interviewees, most of whom are recent U.S. (im)migrants, is the challenges in their daily mobility. But I had never thought of my own challenges and immobilities. To me as a student from China navigating my own path in both personal and academic worlds, border and citizenships are my barriers; ". . . the border exists not merely as a line on a map" (Lunstrum,

2014, p. 104). The border translates into individual bodies as an embodied experience in everyday life. I am always reminded how different I am from the people with a green card or U.S. citizenship, how much more effort I need to make to realize my career goals because of both the physical and social immobility I face.

Wei Li—Shaolu's recent stories make me reflect on my own 27-year journey in North American academia, the shifting from immobility to mobility through the meanings and realities of legal status, citizenship, and nativity.

As a J-1 exchange scholar—Merely two months after my arrival in the United States, UNL's International Office organized a nine-day seven-city across-the-U.S. tour for international students and scholars at the cost of $200. Being a geographer and wanting to explore the United States whenever I could, I got myself placed on the waiting list as soon as I learned about the opportunity but was not cleared to join until a few days before departure—imagine how excited I was! That excitement did not last too long before reality hit on day two of the trip. En route between Chicago and Niagara Falls while inside the group van, a university official pulled out a sheet of paper and announced that "despite that Niagara Falls is so much prettier on the Canadian side, if you are a citizen of any of the countries on this list that I am to read, unfortunately you are not permitted to enter Canada and will have to stay on the U.S.-side instead." He then started to read the names of all countries/areas on the list, of course China was on it, along with Iran, North Korea, Taiwan, among others. He later explained the reasons were not U.S. regulations but Canadian because at that time, anyone who arrived on Canadian soil could self-declare as an asylee and seek asylum. To prevent that from happening, the Canadian government prohibited citizens of "politically unstable countries" from entering her territory without a valid visa. That night, almost half of us stood on the U.S. side of the bridge and waved to the lucky ones as they proudly crossed the bridge and went to the Canadian side to witness the wonder of the beautiful Niagara Falls at night. That was the first-ever reality check for me of *the immobility as a PRC citizen, regardless of my human capital level and despite having a valid U.S. visa.*

As a U.S. Green Card holder—Eleven years passed after that incident and I was an Assistant Professor at the University of Connecticut. A distant Singapore relative was celebrating his 80th birthday, and members of my extended family from China, Canada, and the United States were to join the event for his big milestone in life. I was thrilled to have the opportunity to visit a country with an ethnic Chinese majority as I had by then specialized in Chinese migration as the focus of my research endeavors. I picked up the phone, called the Singaporean Embassy, and inquired about the procedures for a PRC citizen to apply for a short-term visitor visa to enter Singapore. "In which country do you hold your citizenship?" a deep male voice asked at the other end of the phone line.

"The People's Republic of China," I said. "Sorry, we do not offer visitor visas for PRC citizens without a hard copy of an invitation letter from a Singaporean citizen and his/her bank statement," he replied. "But I am a U.S. Green Card holder and have a tenure-track position as an Assistant Professor here with stable income. And I have no intention to migrate to another country. Is it okay if I provide my own U.S. bank statement?" A firm "No" answer ended the conversation. I did not want to bug an eighty-year-old man whom I'd never met before, and even if he'd be willing to provide both required documents, the mail time plus visa application/approval period would render my trip purposeless as I would have missed the celebration anyway. I did not visit Singapore until 2005. Again, I realized my *immobility as a PRC citizen regardless of my human capital level and my U.S. Legal Permanent Resident (green card holder) status.*

As a naturalized U.S. citizen—For a variety of reasons, not the least to improve my transnational mobility as I increasingly participated in international conferences, I decided to apply and was subsequently approved to become a naturalized U.S. citizen in the year 2000. Since then my mobility has increased dramatically, as most countries either do not require an entry/exit visa for U.S. citizens, or a visa can be issued on-site once landed in a foreign country. Exceptions include my homeland, the PRC, which does require a visa for me to go back anywhere in China. So I thought by being a U.S. citizen, my mobility problems were largely, if not all, resolved until the summer of 2005, when I was to visit Australia the first time for a conference. Australia requires U.S. citizens to apply online for an electronic visa and pay a $20 application fee at the time. "That's easy," I told myself and started entering all the needed information online right away. All went smoothly until the question "Place of Birth" appeared. I entered "Beijing, P.R. of China," when a problem occurred. The online system indicated that my application for an entry visa to Australia could not be handled by the system but would have to go through a personal evaluation by Australian immigration officers. "All because I was born in China," I immediately realized as the reason. A few days later my application was approved, but that experience left a scar on my heart. Exactly a year later, I was to go to Australia again for an International Geographical Union's regional conference. With the previous year's experience (i.e., I returned to the United States and didn't overstay my Australian visa), I did not anticipate the same problem. However, much to my dismay, exactly the same problem occurred when my online application was interrupted as soon as the system saw "Beijing, P.R. of China" as my birthplace. These two experiences made me extremely uncomfortable. Should I say it was an "emotional immobility" perhaps? I felt I was treated by the Australian government as a second-class U.S. citizen because of my birthplace—something beyond my control. Despite some unpleasant incidents inside the United States (Li 2005; Li and Yu 2011), I have so far never felt anyone treat me as such in the

United States. Therefore, in answering a recent survey among the PRC-born faculty members in the United States, when a question asked if we'd consider moving to another country, I entered "No" right away to Australia as a possible destination. This incident with the Australian visa application, along with another similar case, made me realize my *continuous inconvenient mobility even as a naturalized U.S. citizen regardless of my human capital level and citizenship status.* What mattered is my nativity of not being born in the United States in the minds of some foreign governments. Compared to the transnational elites, "middling transnational actors" (Ho 2009) do not have a truly footloose mobility, but are more confined by immigration regulations, and the socioeconomic and political conditions (Conradson and Latham 2005; Smith 2005). Our own personal experiences indicate the limitation of physical mobility even for highly educated/skilled legal migrants due to our legal status, citizenship, and nativity.

CHANGING "SENSE OF HOME" AS TRANSNATIONAL GEOGRAPHERS

As transnational geographers, we are mobile and immobile at the same time, or at different times, confined by many factors beyond our own control. But we can at least control our own feelings of where home is that anchor our emotions. Or can we? Here again we found, two decades and one generation apart, that we share similar feelings and emotions.

Wei Li—I did not start questioning where home was for me until the summer of 1992, four years after I arrived in the United States, when I paid my first home visit to Beijing. I was born and raised in Beijing, and I lived there for the majority of the years I lived in China. My father still lived in Beijing and there was no doubt it was the only home I ever knew and had.

That homecoming deeply shook that belief of mine. Beijing changed too fast. Old buildings were demolished, making room for new skyscrapers that sprouted all over. I could hardly recognize many parts of the city that I used to know so well. So much so that I was once inside a cab when the driver questioned me: "Hey lady, you sound like a true local, but how come you don't know any places?" (I turned back to my local Beijing dialect each and every time I was inside a cab in Beijing so people would not detect that I was visiting from a foreign country.) I paused for a second, then I told him a white lie (more true than four years ago though) that "I am a nerdy professor working at Peking University and lived at Tsinghua University, so I rode my bicycle in between the two campuses most of the time." The driver was happy with my answer. Beyond the change in physical appearance, Beijing and my own personal life both went through a turbulent time. I felt the Beijing I came back to was no longer the hometown I left behind four years ago. I was deeply troubled and worried. Therefore, my sole aim was to dutifully accompany my father (as the only child of my parents) in order to make him happy during the 21-day trip. He was indeed convinced and proudly announced to his friends that "my daughter has not changed a bit, albeit in the U.S. for four years!" During the entire time of my home visit, I was eager to return to Los Angeles, where I was studying toward my PhD degree in geography, to resume my inner peace of mind. Or so I thought.

It turned out that returning to Los Angeles, California, brought even deeper turbulence in my mind. I realized it too was *not* my home—I was simply passing through there during a long academic and life journey. I had no clue where I would eventually land, so Los Angeles was a temporary place for me during my study period. Then where *is* home? I had no idea or control over it. I felt completely lost without a sense of belonging—my hometown no longer made me feel at home and where I lived didn't either. I shared those feelings at my first-ever Chinese American studies conference with a renowned scholar and did not get a satisfactory answer. I then shared my feelings with a good friend from our Peking University days, who questioned the validity of my puzzle. "Why do you have to have a sense of belonging?" he asked. It wasn't until years later, that he too had similar feelings during his own first home visit to China.

The sense of belonging was still a puzzle and I did not know where home really was during the entire time I was in Los Angeles. After a couple more visits to China, I continued a transnational life and retained transnational connections between my home country and the United States, where I studied. In February 1997, I defended my dissertation on the Chinese immigrant communities in suburban Los Angeles's San Gabriel Valley. A few days after my defense, I went to the University of Connecticut (UConn) for a job interview for an Assistant Professor position to be possibly jointly appointed by its Asian American Studies Institute and Geography Department. I was thrilled by this academic position that fell exactly where I had envisioned myself to be within American academia. Despite the hospitality of the faculty, staff, and students in both units, I was stunned by the university's rural and almost isolated location. I remember that although I was a trained geographer, in that pre-GPS/Google era it took me at least five minutes to locate Storrs, where the UConn campus is located, in an atlas. "This is totally in the middle of nowhere," I told myself. The on-campus interview confirmed my suspicion—the campus is in the middle of the woods without a normal "American university town" feeling at all. An approximately ninety-yard-long, two-story-high commercial building that included a photo shop, a dry cleaner, a diner, a Chinese buffet, and an independent pharmacy composed what is known as "downtown Storrs." As my then future colleague Fe Delos-Santos told her visitors before approaching that commercial strip not to blink, as otherwise they would miss it. Out of pure curiosity, I asked where faculty members lived. My hosts were impressed with the question and interpreted it as my seriousness about the position and place. A faculty member enthusiastically gave me a brief tour of the adjacent areas and pointed out a few houses owned by my future colleagues. "We don't have a public sewage system so every house digs a well for water and uses a septic tank to treat waste," he said. "What? Am I going to spend possibly the rest of my career in such a place?" I still remembered the sad feelings I had when I first passed by a couple of burial grounds on the roadside, and told myself "I do not want to end up sleeping and becoming a tombstone there." I felt puzzled and thought if I landed there I'd again become a "sent down youth" (an expression used during the Cultural Revolution years in China when Mao sent millions of urban youths to rural areas for farm work) all over again, only this time on foreign soil and not in a Chinese rural area.

After the interview, I had such mixed feelings. I liked the job and the people very much but disliked the place to an equal degree. After boarding my return flight, I started to browse the in-flight magazine as I always do. An article by a travel journalist caught my attention when he wrote something like, "Where is home? Home is not a place or a house. Home is where people you love/care about live and who love/care about you live." "Wow," I thought. "This resolved the puzzle I had for five years!" When the plane finally landed at LAX, for the first time during the six years I was in Los Angeles I told myself, "Los Angeles *is* my home." That emotional attachment and tie to Los Angeles lasted for five more years after I took the job at UConn and even during the first summer after I joined Arizona State University (ASU). I kept part of my research in the LA area and returned there every summer. I never owned a home nor had family in LA, but still, it became home for me only when I was about to leave and had left there.

Shaolu—"Where is my home?" This question has always been an ambiguous one that triggers profound nostalgic melancholy in me since the first day of arrival in the United States. When asking this question among my interviewees during my dissertation research, the first reaction was mostly sighing and pausing. I completely understand where this sadness comes from. I have never doubted my identity as a Chinese. I am deeply proud of China's history, heritage, culture, and its rising economic status in the world. I am also deeply concerned for its issues such as government corruption, environmental pollution, and occasional natural disasters. In my classroom, I let my students know that China is more than just Kung Fu, dumplings, and pandas. But the firm identity as a Chinese was challenged three years ago when I went back to China to visit my friends and family, which was the first and only time I visited after two years of staying in the United States.

I took the airplane to Beijing and the high-speed train to my hometown Qingdao. That way I could visit my friends and mentors in Beijing and take the same train line as I used to take when I was a master's student in Beijing. My home in Qingdao is only a ten-minute walk from the train station. It was the same route I walked for seven years of being away from home. But this time I was lost. Qingdao is a large coastal city with eight million people. To compete with other cities in China and in the world in the global era, it has been going through dramatic changes, reflected in the construction of new infrastructure and buildings. It is home to the longest overwater bridge in the world. The starting point of the bridge was right in my neighborhood, which used to be the playground of me and my childhood friends. The year I left home, the bridge was under construction. The year I went back home it was already completed. The construction of the bridge completely changed the urban structure and street systems of my neighborhood. The small alleys where I used to walk and run were wiped out, giving way to the broad avenues that lead the traffic to the bridge. The historic houses were replaced by high-rise loft apartments and office buildings. Luckily my apartment building was saved but it was dwarfed by the surrounding giants, blocking my ocean view which used to be my shelter during my childhood.

Where is my home? I realized *that* home I have missed so long only exists in my memory now. This sense of loss of home is followed by a chain of reverse cultural shocks. The first shock was environmental pollution; Qingdao used to be famous for its clean air and pleasant environment by the ocean, attracting thousands of tourists each year. But this image was challenged by the environmental problems caused by economic development and the increasing car ownership. It is good that people's living standard has improved, but at the expense of clean air and blue sky. The sky was covered by smog almost nine out of ten days during my visit. The scenery of beautiful stars and moon over the horizon of the ocean also became my everlasting memory. The second shock was transportation. Obtaining my driver's license in the United States made me cautious of sharing roads with bikers and pedestrians by giving them priority. But I found it a difficult and risky task to cross the street in cities in China. It took me a while to get used to the fact that pedestrians should yield to vehicles, otherwise I would either be honked at or smashed. The third shock was the difficulties of being a single, highly educated female in her late 20s, despite the changing gender dynamics and improving women's status in modern China. The pressures from family and friends to persuade me into dating and marriage almost drowned me. This also made me appreciate the freedom in the United States of singleness, of pursuing my career without the burden of the expected role of a Chinese woman, without hurrying into a marriage just for the sake of marriage.

I was looking outside the window of my room at the street where I used to expect my parents to come back home from work when I was a child, at the bus stop where I waited for the bus to school when I was a teenager, at the ocean where I watched the beautiful sunset and waved to the ships passing by. The smell coming from the kitchen of each home lingered in the neighborhood, mixed with the odor of seaweed and the chanting of seagulls, and greeted the commuters coming home from work. I asked myself, is this my home? All these peaceful scenes were so familiar, yet so strange; they are so close, yet so distant. I felt I was no longer involved in the scene, but rather I was looking at a painting; I was an audience, an observer, a tourist, an outsider.

Realizing I will depart again and go back to where my routine is in the United States, I feel sad but at the same time so relieved. The ambiguous and conflicting feelings struck me. I miss home, but where is my home? My identity is clearly "Chinese," but China is no longer my home in reality and only my home in memory. Both of us, me and this place, have changed. The place I return to is no longer that place I left behind, this person is no longer that person I used to be. I still experience the place, but less "from inside out" than "from outside in" (Marcus 2011, 22), through the transnational connections with my parents and my friends. I no longer take for granted the things I used to see and experience in my childhood. I do not belong there despite that this is where my family lives and where my childhood was. But is the United States my home? I did miss my office, the beautiful lawn where I sit and chat with friends, the hill where I run every day and visit cows, sheep, horses, trees, and stone walls. That is where my day-to-day life is. That is where my dream career and my possible future family is. But I am a Chinese citizen, on an F-1 student visa, traveling with my passport wherever I go, still trying to learn American culture and speak good English. I have never felt fully integrated into the American culture, and I am not sure I will or want to be. Thinking of these, I was overwhelmed by a sense of rootlessness and homelessness, strong enough to make me fear. Looking at my two large suitcases which had been quietly sitting in the corner of my room, which are the exact same suitcases I carried with me two years before when I started the U.S. journey, I realized *that* is my home. As Anzaldua puts it (also quoted in Swanson 2014, 27): "I am a turtle, wherever I go, I carry 'home' on my back" (1999, 43). Now that I think of home again, I ask myself, am I really home-less/place-less? Chinese culture values roots. Chinese culture believes in roots, reflected in the old saying "Fallen leaves return to the roots" (落叶归根, *luo ye gui gen*). Paradoxically, the Chinese are also cosmopolitan and mobile, reflected in another popular saying: "Feel at home wherever one goes" (四海为家, *si hai wei jia*). The Chinese have been one of the most uprooted and adventurous people, scattering all over the world. I opt for the second idea. I might experience temporal rootlessness but that does not mean homelessness; rather, my home has been multiplied. Similar to Wei's sense of home in Los Angeles when she was about to leave for the University of Connecticut, it is only now when I am about to leave the University of Connecticut that I started to realize I have become so attached to it over the years. It is my home, despite how strange and placeless I felt five years ago when I first came here. As an international scholar in a transnational era, it is hard to claim a single home and the meaning of home as to *dwelling* and place of *being* (Heidegger 1962). Home-ness has long been examined in dualism between home and homelessness. In a transnational world, it is a spectrum. Sense of home is not fixed or must be bound to a particular place. It is to be established and re-established. The attachment to place increases with the length of stay. It becomes even stronger while being away from home, both physically and temporally. The temporal dimension of home should be valued as much as the spatial dimension; what *used to be* home in the past does not make it less meaningful or legible than what *is* the current home or what *will be* the future home. *That* home is remembered and sometimes dreamt. Homes in different time periods are all unique and profound. It is not one replacing another. They are all part of me and I am part of them. They are all written in me and my trajectories.

What both homes have left me is the unique transnational experience and a critical perspective on both countries, which I would had never developed if not for the several years of studying in the United States. This perspective allows me to look at two countries from a comparative perspective, be it natural resources, politics, society, economics, ideology, people, or culture. It is most valuable for a geographer, as it enriches my teaching and my research on transnationalism and mobility. My dissertation in Flushing benefits from the transnational connections which help me expand my social circle and establish friendship with my interviewees. This gives me the privilege of interrogating the interrelations among transnationalism, mobility, place, and scale, through experiencing them as an insider. What I took for granted in both countries began to appear to me as very interesting research topics. I share this experience with my students so that they can look at how America differs from other contexts with a comparative angle. This transnational relocation experience will continue to enrich and enhance my life experience both personally and professionally.

So the two authors, exactly two decades apart (1992 in Wei's case versus 2012 for Shaolu), shared the same feelings that their respective hometowns had changed too fast during the few years after they left; and the places where they studied in the United States could hardly be called their home. The combination of these feelings deeply shook their sense of belonging that could have anchored their life in a foreign land. In this accelerated globalization era, no place or person is static. Would the changing of a place contribute to one's identity and sense of belonging? We will address this in the next section.

CHANGING OF A TRANSNATIONAL PLACE THROUGH TIME

As the readers may have noticed, the two coauthors either studied or worked at UConn, respectively, but more than thirteen years apart (1997 and 2010). We now reflect from a micro-environment perspective on how a specific institution changes over time and over two generations.

Wei Li—As indicated in the previous section, my initial impressions of Storrs were not positive ones. But the physical isolation *per se* was not the major concern. It was more about the human environment. To this day, I clearly recall my first-ever breakfast at a Friendly's diner on the second floor of that ninety-yard-long commercial building. The then-director of the Asian American Studies Institute kindly opened the door for me, so I could walk into the restaurant before him. I am not exaggerating here, but more than half of the mostly white, elderly customers there when I entered stopped eating completely, turned their heads around, and looked at me instead. I had an extremely uncomfortable, "sunk to stomach" type of feeling. "I am *not* an animal in a zoo to be displayed for show purposes," I thought to myself. I had lived in Los Angeles for six years by then and two and a half years in the Washington, DC area. I had been used to living in diverse transnational communities where people who looked different and belonged to various racial/ethnic/cultural groups were the norm. This was not so at Storrs, Connecticut, at the time. There was a small number of very vocal and politically active Asian American faculty, staff, and students, and some Asian international students. Obviously, outside the UConn campus, the majority-white local population was not used to having Asian faces around in their normal daily settings. Still, lured by the good professional development opportunities at a Research I institution, I nonetheless signed the offer letter and moved to Connecticut in the summer of 1997. Before that move, I called the director of the Asian American Cultural Center, asking where I should seek a place to live. She said, "There are two types of housing you may choose: the Chinese 'ghetto'—off-campus graduate housing where most Chinese graduate students congregated; or the junior-faculty 'ghetto'—an apartment complex on Foster Drive, seven miles south of campus, where a large number of UConn new hires live." "I will find one at the junior-faculty ghetto," I said without hesitation. I was hoping my American working and living experiences would enrich my life with multiracial and multicultural interactions. The junior-faculty ghetto complex was indeed diverse, including the family of an African-immigrant colleague in the geography department, which slightly eased my uncomfortable feelings.

A few months later, I was invited to participate in an NSF-funded workshop on "Race and Geography" at the University of Kentucky campus at Lexington. Among the attendees, there was a Korean American professor from a University of California campus. After a couple of meals outside the campus, she pulled me away from the crowd. She asked, "Did you notice people here stared at us two? I felt really uncomfortable and just want to go home soon." I told her I was already "battle-tested" and no longer even noticed that, let alone felt anything, as I was used to being stared at in Storrs. So people have to adjust to their surroundings by internalizing their feelings and move on. Or do they? How does a place or an institution also change over time? How do people perceive the same place or institution after some years in this accelerated globalization age? Let's see what Shaolu's initial impressions/reactions were when she arrived at UConn in 2010—how were they similar to or different from mine?

Shaolu—Compared to 1997 when Wei was at UConn as a faculty member, the department, the campus, and the surrounding environment have changed dramatically. The year I came to UConn was in 2010. I have been asked several times the reasons I chose UConn. First, ranking. Second, location: between Boston and New York City. As an urban geographer, being close to cities is a must if not living in one. Third, diversity: the graduate students and faculty, including research interests in nationality, race, and gender. As an international student, I want to make sure that I am in a welcoming, open, and diverse environment.

The first night I arrived at UConn still feels like it was yesterday. I can still remember my excitement, surprises, and fear. I first landed at the BDL airport in Hartford. A senior Chinese student whom I met through the UConn Chinese Students and Scholars Association (CSSA) offered me a ride from the airport to the campus. CSSA is a student organization where students meet and help each other, such as arranging rides and accommodations for the newcomers. The association is supported through the platforms of a mailing list and QQ (Chinese Skype) group where the Chinese students at UConn share information and experiences of studying abroad. The CSSA helped me through my first difficult year at UConn while adjusting to changes both in life and in academia.

Similar to other first-year students, I also went through cultural shocks. The first shock was the geographical change, not only across the ocean and country borders but also from urban to rural setting. Even though between two large metropolitan areas, Storrs is in the "quiet corner" of Connecticut with a rural landscape, as Wei observed thirteen years prior to when I landed, which was out of my expectations. Growing up in a city of 8 million and going to school in Beijing with a population of 13 million, I felt completely disappointed the moment I landed in Hartford. The 40-minute drive from the airport to the campus was nothing but highways, woods, farms, and scattered houses. Mixed with the feeling of homesickness, I asked myself, "Is this the place where I want to spend my next five years?" I was struck by a sense of placelessness. The second shock was transportation and housing. In Storrs, no car means *no legs*. As a Chinese student spoiled by the sound public transportation in cities, the first cultural adjustment is to learn how to drive. Fortunately, thanks again to the Chinese community on campus, I obtained my driver's license and bought my first car after one year. So the second year, I was able to move off campus to Foster Drive, coincidentally the same complex and neighborhood where Wei lived 1997–2001. However, it was no longer a "junior faculty ghetto," but an "international student ghetto." Foster Drive has become a popular residential area for the graduate students for its relative closeness to Willimantic (the largest town within a ten-mile radius) and the campus, with a bus serving in between.

Compared to 1997 when Wei first arrived at UConn as a faculty member, the department and the whole university was much more diverse in 2010 when I was admitted, although UConn was not as popular among Chinese students as those like USC where Wei obtained her PhD degree (University of Southern California; it even has the nickname "United States of Chinese" due to the high presence of Chinese students). The department then was small but cozy. Among ten graduate students, three were Chinese. I was lucky to be one of the three doctoral students who were admitted in 2010. The other two were from Sri Lanka and Russia, both female students. Now the department has become greatly enlarged and diversified. Among thirty-two graduate students, eighteen are international, eleven from China. The percentage of international students in the university has doubled from 5 percent in 1997 to 10 percent in 2014 (UConn OIR, 2015). The faculty has also diversified. The percentage of international faculty has also doubled, from 1.8 percent to 3.7 percent (UConn OIR, 1997, 2014). The internationalization of the students has mostly happened in the past five years during my stay. The number of students from Asia, especially from China, has skyrocketed.

The diversity on campus also impacts the surrounding community, as international presence has dramatically increased. The local residents and businesses have also become more used to the international environment. This changed the economic and cultural landscapes around campus. As more Chinese students come to UConn, more Chinese businesses such as restaurants and groceries pop up here and there. The environment around the university is welcoming, but whenever I am exposed to a white-dominated environment farther from the campus, I still experience the similar situations as Wei did almost twenty years ago. Being a minority I did feel the self-consciousness and discomfort of being different and even being alienated. I hear about the stereotypes of Asians, especially of Asian women. For example, the stereotypes are often heard in daily language that Asian men are reserved, obedient, and rude, and Asian women are hypersexual and bad drivers. When I was a patron at a local restaurant or a bar, being stared at or treated differently from my American friends was not uncommon. Many of the unpleasant encounters are subtle in people's facial expressions and tone of speaking. This becomes more obvious to me when I experience the contrast while doing fieldwork every summer in New York City, a more ethnically and culturally diverse setting, similar to Wei's case when she lived in DC and LA.

Postscript—from June 24–27, 2015, Wei finally had a chance to revisit UConn, while participating in AAG's departmental leadership workshop, after fourteen years. Shaolu showed her around the rather impressive new "downtown Storrs," which has a somewhat decent "college town" feeling. The old short strip is remodeled and dwarfed by several condo-apartment buildings with retail chains such as 7-Eleven and CVS, as well as a grocery store and a bank, among other services. As a visitor, Wei enjoyed seeing the wooded New England scenery and campus development. However, the evening when Fe, Shaolu, and Wei went to the upscale Altnaveigh Inn restaurant, the old feelings of being stared at, even maybe by friendly eyes, came right back, as she noticed she was stared at again.

The different but also similar experiences of us, the two generations of Chinese geographers, in the same micro-environment over the past two decades demonstrate that despite the increasing diversity in a globalization era, something fundamental in U.S. society has not changed much, if at all. Also, places do matter, as they continuously impact people's daily experiences and identities. Equality and equity remain a goal hard to achieve for many, including those with socioeconomic privileges.

CONCLUSION

For Shaolu and Wei whose research areas are immigration and transnationalism, mobility, borders, and home are the forever topics they research as an outsider and simultaneously experience as an insider in their daily life. Two-decade time difference between these two-generation Chinese geographers does reveal differences in their transnational experience both in academic and personal lives. However, some experience remains unchanged: limited physical mobility which further constrains academic mobility, and an ambiguous and complex sense of home, as well as the impacts on them by the environment at the institutional level and the surrounding community.

REFERENCES

Anzaldua, Gloria. 1999. *Borderlands/La Frontera*. San Francisco: Aunt Lute Books.

Bauman, Zygmunt. 2000. *Liquid Modernity*. Cambridge, UK; Malden, MA: Polity Press.

Conradson, David, and Alan Latham. 2005. "Friendship, Networks and Transnationality in a World City: Antipodean Transmigrants in London." *Journal of Ethnic and Migration Studies* 31(2): 287–305.

Elizabeth Lunstrum, 2015. "Borders, States, Nations: Living Political Geography." In Tremblay, Rémy, and Susan Wiley Hardwick, eds. 2015. *Transnational Borders,*

Transnational Lives: Academic Mobility at the Borderland. Quebec, Canada: Presses de l'Université du Québec.

Hamilton, Gary G. 1999. *Cosmopolitan Capitalists: Hong Kong and the Chinese Diaspora at the End of the 20th Century*. Seattle, WA: University of Washington Press.

Heidegger, Martin. 1962. *Being and Time*. New York, NY: Harper Perennial Modern Classics.

Ho, Elaine Lynn-Ee. 2011. "Migration Trajectories of 'Highly Skilled' Middling Transnationals: Singaporean Transmigrants in London." *Population, Space and Place* 17(1): 116–129.

Li, Wei. 2005. "U.S. Immigration and Racialized Assimilation" 57–65 in John Frazier and Eugene Tettey-Fio, eds. *Race, Ethnicity, and Place in a Changing America*. Binghamton, NY: Global Academic Publishing

Li, Wei, and Wan Yu. 2011. "Racialized Assimilation? Globalization, Transnational Connections, and US Immigration" revised/updated chapter, 33–44, in *Race, Ethnicity, and Place in a Changing America*, 2nd edition, John Frazier, Eugene Tettey-Fio, and Noah Henry eds. Syracuse, NY: Syracuse University Press.

Marcus, Alan P. 2011. "Transnational Rio de Janeiro: (Re)vistiting Geographical Experiences." In Friedman, May, and Silvia Schultermandl, eds. 2011. *Growing Up Transnational: Identity and Kinship in a Global Era*. Toronto: University of Toronto Press, Scholarly Publishing Division.

Office of Institutional Research and Effectiveness. 1997, 2010, 2014. http://www.oir.uconn.edu/. Last accessed, July 2, 2015.

Ong, Aihwa. 1999. *Flexible Citizenship: The Cultural Logics of Transnationality*. Durham: Duke University Press Books.

Smith, Michael Peter. 2005. "Transnational Urbanism Revisited." *Journal of Ethnic and Migration Studies* 31(2): 235–244.

Swanson, Kate. 2015. "Where is Home? An Autoethnography of Academic Migration." In Tremblay, Rémy, and Susan Wiley Hardwick, eds. 2015. *Transnational Borders, Transnational Lives: Academic Mobility at the Borderland*. Quebec, Canada: Presses de l'Université du Québec.

Tsuda, Takeyuki. 2011. "When Human Capital Does Not Matter: Local Contexts of Reception and Immigrant Wages in Japan." *GeoJournal* 76(6): 641–659.

NOTES

1. The authors thank editor of this volume, Alan P. Marcus, for the invitation to contribute, which provided us a unique opportunity to reflect upon our work and lives in the United States. We would also like to thank Fe Delos-Santos of UConn's Asian American Studies Institute for the nice conversation that sparked and connected our memories and ideas of UConn and Storrs that contribute to the chapter. We are also grateful for Ted Knowles of Simmons College who edited and proofread the first draft.

2. In order to get admitted to most graduate schools in the United States, students from outside the United States are required to take the GRE and TOEFL. TOEFL stands for Test of English as a Foreign Language. It is a standardized test of English proficiency for non-native English speakers. GRE stands for the Graduate Record Examination. It is a standardized test required for admission to graduate schools in the United States.

3. Schengen visa is the visa that covers the entire Schengen Area, including most of the European countries. The applicant for Schengen visa needs to have a valid visa to the country he/she would return to (http://www.schengenvisainfo.com/).

4. OPT, Optional Practical Training, is a period when undergraduate and graduate students from outside the United States are permitted to work in the United States for a period of time after or near completing their degree. The permitted time is usually 12

months. Students from STEM are eligible to extend their OPT period (http://www.uscis.gov/eir/visa-guide/f-1-opt-optional-practical-training/f-1-optional-practical-training-opt).

FIVE

Polish(ing) My Identity

Weronika A. Kusek

I started writing this chapter while sitting on a beach in St. Thomas, enjoying the sun and the ocean breeze after enduring five long months of harsh winter weather and a couple of additional post-winter transition months in the Upper Peninsula ("UP") of Northern Michigan. Now, Marquette, Michigan, has become my new home after I accepted the offer of a tenure track position at Northern Michigan University in 2014. I was thinking about my first year in the UP, and how Marquette will likely be my home for an unforeseeable amount of time, which means years of long winters but also years of growing into a welcoming, friendly, and professional community.

I was born and raised in Poznan, Poland. I had to ask myself: How and when was this path written? What lies ahead? After nine years of attending college, graduate school, and completing a PhD in the United States, I moved to Marquette, MI—a town of 19,000 people—a place I did not know existed when I was enrolled in my first geography classes in Poznan, Poland. I will explore these questions in this chapter.

PLACE

Place has been a subject of research for centuries. A place can serve as a platform for collecting experiences, meanings, and connections to other people. As I transition through locations, place for me becomes the combination of features and factors of the external with the ideas and activities that I bring with me. Place is something I experience, but also something I recreate and practice. This interaction between the internal and external allows me to believe that cumulatively and chronologically, places have shaped who I am. Places made me a geographer, a transnational migrant, a researcher, a college professor, a friend, a wife, a daughter, and a granddaughter.

In this chapter, I present how places have shaped my personal and professional lives and how they affected my migratory journey. I build upon and reflect on geographic concepts such as place, community, migration, transnationalism, and sense of place as main concepts used to create what Alan P. Marcus has called an "autobiogeography" (see introduction in this volume).

I was always a geographer at heart. As a young child, I remember sitting on my bedroom floor with a big geography atlas, looking at the map of the world, marking places where my relatives lived. I remember learning about Chicago, Illinois, where my uncle had moved to during political unrest in Poland, trying to imagine what the place looked like.

Place as home has been a traditional area of study of geographers (e.g., Tuan 1991; Seamon 1979). Tim Cresswell (2004) states that home is the place where an individual feels a sense of attachment and rootedness. Gaston Bachelard (1994) believes that home determines how we understand other places. So how did Poznan, Poland, predispose me for my migratory journey?

Although no member of my closest family was a migrant, my immediate family loved to travel. Born in 1983, I was fortunate to travel with my parents throughout Europe as a child in the 1980s and 1990s. I visited Spain, Italy, Greece, Austria, and many other European places which exposed me to the joy of travel and the experience of new places, cultures, and ways of life. When thinking about these experiences, however, I realize that they set the foundation for my interest in geography, but I am not sure that they had any significance in my journey as a migrant and transnational geographer. I never thought that one day I would be reflecting on a ten-year stay in the United States. Instead, for many years, the most significant place in my life was Poznan, Poland—my "home"—and a place that I thought was going to be my home forever.

POZNAN, POLAND—PLACE OF CHILDHOOD HOME

At the time of my birth, Poznan was a place characterized by a communist reality that manifested itself in a high police presence, empty store shelves, high level of social frustration, and feeling of isolation. Traveling to Western European countries was restricted. The chief characteristics of that place changed dramatically with the fall of communism. My parents, like many other Poles, welcomed democracy. As pioneering business owners in the restaurant and hotel industry they were also able to take full advantage of new travel opportunities, taking me and my brother abroad, often twice a year. But once back home, my life was focused and defined by Poznan—a city of half a million people that was also one of the hubs for defining the new and future reality of Poland. Thanks to its geographic proximity to Western Europe, the city always attracted visitors from Germany and other Western countries for tourism, business, and family reasons. Poznan was thus a place which, for the standards of Poland in the early 1990s, offered relatively more exposure to diversity than other cities. Poznan was quickly becoming a place with good schools, prestigious universities, delightful culinary experiences, modern shopping, and much more.

My teenage years overlapped with Poland's political and economic transition into the European Union. Unable to distinguish whether my broader perspectives were a result of my age or Poland's increasing international exposure, as a teenager I felt a member of a bigger community of other citizens of European urban areas enjoying a vibrant, walkable city. I observed the city and its people on my way home from school—on a crowded tram—noticing new venues, stores, fashions, languages. I had the opportunity to interact with people from Poland as well as tourists who were visiting Poznan and stayed at the hotel that was my family's business. During my high school years, I helped at the hotel as a receptionist, bartender, or translator. I was also encouraged and supported by my family to attend a private high school where the promise was to educate global citizens. This all resulted in the fact that at the end of high school, I thought I was well traveled, spoke English, and after spending seventeen years in Poznan I was ready to experience the world in a less pre-packaged sense—to date defined by tourist experiences with my parents, interactions with international visitors in my city, and international media. I knew there were other places in the world I wanted to learn about and experience. At the time, I just did not know how to reach these places. Thus, when my high school was looking for a family to host an American exchange student I was the first person to volunteer. Ashley, an African American teenager from Ohio, stayed at my Polish home for a year. This was my first real exposure to "American culture." During that experience I also learned that learning about the American culture also meant recognizing African American culture. While I am still learning what "African American" culture means, thanks to Ashley, I learned about the complex identities composing the broader term "American." Ashley helped me to gain new perspectives on Poznan as my home and as a place, since I was seeing it through her eyes. But I also learned about the United States as a place from perspectives I never knew existed. Ashley shared with me her American experience and the issues she believed were troubling her country—something I, as a young, white, Polish teenager would have never had an opportunity to learn at a Polish school

After Ashley had returned back to her home in the United States, within a couple of years I was reminiscing about these experiences while boarding an airplane to study abroad in Toledo, Ohio. I was traveling to attend a semester of college, to experience the American culture first-hand, to polish my English (no pun intended), and to visit places I had learned about in my childhood geography atlas. At that time and for much longer after that, I did not know I was beginning a migratory journey and a transnational life.

MIGRATION AND TRANSNATIONALISM

Migration is not a new experience for Poles (Kusek 2015). In fact, the Polish diaspora is one of the largest in the world. The Polish Ministry of Foreign Affairs estimates that up to 20 million people of Polish ancestry live outside of Poland. Historically, Poles were migrating to the United States, Canada, and Australia, and have recently re-energized their migration to the United Kingdom. In fact, the United Kingdom has welcomed more than 600,000 Polish migrants since 2004 (Kusek 2014), and the United States is home to several million people of Polish ancestry. Due to its turbulent, violent, and uncertain history, Poles were often forced to leave their homeland due to political oppression and long periods of economic distress. Even recently, after 25 years of successful democratic and economic transitions, the wage gap between Poland and pre-expansion EU countries remains significant. Labor migration thus became a choice for many when the European Union expanded in 2004, accepting member states from post-socialist countries of Eastern and Central Europe. Although most Poles chose the United Kingdom as a place to work, study, and live, due to the proximity to Poland, high wages, and lack of employment restrictions for EU members, I decided to move to the United States. As I mentioned before, however, my decision to migrate never materialized in a single event. In fact, after living in the United States for many years I still did not consider myself a migrant. I rather saw myself as an international student, a member of a global community, or a rising "Eurostar" (Favell 2008; Kusek 2015).

For a long time, the word "migrant" had a negative connotation in my mind. To me, "migrant" was a label that, in popular media especially in the United States, was used to describe people who need social services, rely on the welfare system, work in various jobs at below-market rates, and struggle with assimilation. While I did not consciously compare myself to "migrants," I never associated myself with that label. As an international student, not only did my life not resemble the one portrayed in the media as the life of "migrants," but the entire American reality, migrant and host community dynamics were all a show in front of my eyes. I was not part of this. I was just observing as an outsider, an international student who was planning to return to Poland soon. Although I had long abandoned my original intent to stay in the United States for one semester, I now intended to leave the United States as soon as I graduated from college. I thus tried to maintain active connections with elements of my past life. I called my parents and family every day thanks to the wonders of Voice over Internet Protocol services. I read and followed up on news from Poland and the European Union. I traveled to Poland for Christmas and welcomed family members in Ohio. My plan was to return to Europe and work for one of the EU institutions. As I think about that mindset from the perspective of what has now become a decade spent in the United States, I missed many signs and indications that my return to Europe was not imminent. While I had planned to return to Europe, I was increasingly becoming part of a network of friends, neighbors, colleagues, and future mentors which I simply never had the chance to develop "back home." Furthermore, I was developing an identity in the United States that would not exist in Europe and one that I was enjoying. In Toledo, Ohio, I was the "friend from Europe," a girl from a country where people's grandparents came from, the girl that had actually visited dream honeymoon destinations. I was also the "white sister" at many of Ashley's family celebrations, and a close friend to a group of very random people who felt amazingly similar and close due to a shared experience of migration. Finally, Toledo, Ohio, and the United States were the place where I was shaping my adult independence—a process that, over time, has allowed me to develop a unique comfort zone of daily experiences and habits that I would find extremely difficult to recreate in Poznan, Poland, or any other place. Due to this development, I stopped "going back" to Poland. At a certain point, I realized that my trips to Poznan were now called "going" to Poland.

I have remained in a mindset which has allowed me to enjoy connections to several *places* but also sometimes makes me feel disconnected from any. In many ways I am American. I think so due to my emotional attachment to things American, as well as my adoption of many American behaviors and cultural norms and habits (if such generalizations are even possible). For example, I understand and become emotionally engaged in local and national politics, I understand American social and behavioral standards, rules, and protocols (even ones which are based on not having any standards or protocols), I follow and actually love many American traditions, and I have a Costco membership. At the same time, I get excited about expressions of Polish culture in the United States, highlight things that celebrate Poland's contemporary successes and growth, and travel to Poland frequently. I bring back and proudly display Polish fashion goods which are not only unavailable here in the United States, but thanks to their modern design and quality allow me to participate in Poland's growth by literally wearing it. However, when I am in Poland I feel I can never return to the mindset of a local who was traveling in a crowded tram looking at changing people and landscapes. Poland now changes too quickly, and the Poland I visit is not the Poland I remember. When I am there, I know I am not American, but I am a lot like an American traveling to Poland.

The most complex thoughts about my identity actually often occur on an airplane and within airports. These journeys make me notice certain behaviors and practices that are common across European countries and different from American ones, thus allowing me to discover my European identity as distinct from my strictly Polish one. At the same time, these same journeys highlight my undeniable connection to the United States. I could call this the "LH443 Frankfurt to Detroit syndrome." I notice and relate to sharply dressed Europeans sipping their morning cappuccino from a ceramic cup, instead of a *grande vanilla latte* from a carryout paper one. I know they don't want to talk to me, and if we make eye contact, they will not nod or smile. I am fine with that. But I also see the family sitting on the floor of the gate reading their Kindles, wearing their "Central Michigan" or "Pittsburgh Steelers" sweatshirts, who smile to me as I sit down at the gate and allow me to feel an immediate connection and affiliation. Finally, to add even more complexity, as a Polish migrant to the United States I am also very oddly connected to a very popular and growing experience of recent Polish migration to the UK, and thus through my disconnection I feel I also participate in the migratory reality shared with many other young Poles. As I will discuss in the following sections, through my research I am provided with the opportunity to discover and explore these overlapping and conflicting identities. The path to that research was shaped by my American education, and that started in Toledo, Ohio.

TOLEDO, OHIO—PLACE OF MATURITY AND INDEPENDENCE

Toledo, Ohio, is a medium-size American midwestern city. One challenged by rust-belt realities, but also benefiting from the ability to use its smaller size to redefine itself and organize around its key institutions. Thus, Toledo is defined by the University of Toledo, two multi-hospital health systems, a minor league baseball stadium, a legacy of manufacturing and automotive companies, and miles of sprawling suburbs. The city is not a major destination for migrants, is a rare destination for international tourists, and is a fairly self-contained universe. When I first arrived in Toledo I did not understand the American system of education. I enrolled in a community college, as it was the least expensive that offered exactly what I needed: English courses for non-native speakers. Although I achieved my goal at this college, the experience was richer than the tangible results of securing more linguistic fluency or familiarizing myself with the American culture. This was also the first time when I began to reflect on my national and ethnic identity. This experience started with a sociology professor who, after learning that I was an international student from Poland, made a comment in front of the whole classroom: "You don't look Polish."

"What does it mean to look Polish?" I thought to myself, and even today I often reflect on this comment. While the professor did not mean it in a demeaning way, the question allowed me to gain insight into the American perspectives on place—as defined by the stories and memories of their ancestors. These were often perceptions of places that no longer existed, and perceptions that defined and still define many conversations for me and my fellow Polish expats.

As opposed to many stories of other international students, during the first months in the United States I was not homesick. As I was counting the days I had left in the United States I was more afraid I did not have enough time to learn all that I wanted. The fact that I arrived in Toledo with my Polish boyfriend, Jacek (my husband now), certainly helped me to overcome any challenges related to psychological, emotional, and physical separation from my family, described by many academics.

Jacek and I were committed to learning as much about the United States as we possibly could (see Kusek and Radomyski 2008). We watched national and local news, read university newspapers, made a lot of effort to meet neighbors, other students, and community members, and attended any and all social activities organized by the university and the college. We also benefited tremendously from Ashley's father's generosity in organizing time for us. Thanks to Tom, we drove and flew to many states, experiencing very diverse perspectives and geographies, and meeting various people who later influenced our paths.

While there are many jokes about Toledo, mainly promoted by people who live in Toledo, this city, in a very positive sense, became the anchor for my life in the United States. During my education I made multiple attempts to return to Poland—after the first semester, after the first year, and after finally completing my Bachelor's degree. With my graduation looming, I am not sure if we were simply lucky or if we truly experienced the fact that merit is rewarded in the United States. Despite our initial plans, I applied and was accepted to a graduate program at the University of Toledo, receiving full financial support, and Jacek completed his graduate education and secured employment at a local hospital. We both found ourselves in positions which called to our American and Polish cultures—we appreciated the opportunity that was presented to us and were not going to let it go to waste.

In light of my future research and my effort to navigate my conflicting and overlapping identities, I realize that graduate school, through teaching assistantships, helped me to develop as a geographer and college teacher, and also allowed me to understand American students from many backgrounds and with many stories. For the first time in my life I was directly connected to suburban privileged students, white underprivileged students, minority underprivileged students, minority privileged students, future professional athletes and future professionals who paid for college through athletics, veterans, non-traditional students, and others. I often felt that the experience and knowledge I had gained through hosting Ashley in Poland, combined with self-reflection, allowed me to approach students with sensitivity that I could be lacking if I had missed out on the opportunity to be a host family for an American student.

In addition to my academic growth, Toledo became the place which set the foundation for major milestones in my personal life. After living together for several years in Toledo, Jacek and I got engaged, and married the following summer. Interestingly, both the engagement and marriage occurred in Poland, but we knew they were results of our daily lives in the United States. At the time we got married, Toledo became a platform for experiences, meanings, and connections between us and the people whom we met during our five-year journey. It became a place where, day by day, I assembled my adult life. In Toledo, I learned how to cook, clean, and fill the role of a wife that was embodied by my mother or grandmothers.

The ability to develop professionally and personally as an adult in Toledo further fueled identity conflicts in my life. Due to my strong connection with my family in Poland, I traveled and still travel to Poland a couple of or more times per year. As Toledo became the stage for my adult life, it became my "home." Yet I used that same word to refer to Poznan and Poland. As a graduate student in Toledo, as I traveled to Poland I often asked myself what I truly considered "home." Did I have one home, two, or maybe none? Poznan was a place where my I met with my grandmothers, who always cried on the day of my departure, grandfathers, parents, brother, in-laws, friends, and cousins, but Toledo was a place where I functioned as an adult. A place where I had a very satisfying professional life, where I supported myself financially, a place where I was a wife, a friend, a woman. Poznan, on the other hand, was a place where I was a child, a cousin, a granddaughter. Thus, after five years in Toledo, it was time for me to pack, leave our local friends behind, and move to Kent, Ohio, to begin a doctoral program in geography at Kent State University.

KENT, OHIO—PLACE OF DEVELOPING MY PROFESSIONAL CAREER

In the fall of 2010, I arrived in Kent, Ohio. Jacek was still working in Toledo on a work visa, which made it fairly unlikely for him to be able to find and secure a job in the Cleveland area. I rented a room in a home owned by a widow who helped international students find housing and often allowed them to stay at her home until they found another reasonable option. It was a place where I could stay for three to four nights a week. Afterwards, every weekend, I returned to Toledo to spend time with my husband.

The biggest challenge I faced during the first semester of my doctoral program resulted from the fact that I did not feel emotionally engaged with the people, the university, and the city where I was studying. My "homes" were in Toledo and in Poznan. As I was attending doctoral classes, teaching my own class of sixty students, and commuting each weekend and sometimes twice a week, I was in a constant move, spatially separated from my family, my husband, and my departmental colleagues.

As a geographer working on her PhD, I knew about the importance of community relations, sense of place, and cultural adaptation. Upon hearing my struggles, researchers would likely place their palm on their forehead because many had agreeably emphasized that community interactions are essential for cultural adaptation. I found myself in a position where I did not belong to any community. Poznan was still my childhood home, my husband whom I missed dearly was in Toledo, and Kent was a place where I wanted to develop my professional career but only enjoyed through the inside of classrooms and my temporary residence.

In the spring semester, my life changed tremendously: my grandmother passed away—an event which affected my sense of belonging. I realized that Poznan is a home as long as my loved ones are there. If my loved ones disappear, I felt my connection to that place would be lost. In that sense, I came to the conclusion that Poznan is a platform for experiences, meanings, and connections generated between me and my family.

In unexplained ways, at the same time Jacek was able to secure a job in Cleveland. We moved to a charming suburb of Cleveland, rented an apartment, and I was finally beginning to feel a member of the Kent community. Today, I can say that the three years I spent in Kent were the best years of my life. I became fully immersed in my department and its activities and culture. I was blessed to work with a very committed advisor who became a true friend, I took interesting classes, I was given the opportunity to teach a variety of classes, I committed to university service, and I conducted international fieldwork. I progressed professionally to the point where I felt satisfied with who I was as a geographer, an academic, a college instructor. I created my identity and I felt a sense of belonging. Jacek and I made friendships, we discovered our favorite places to visit, and we simply achieved a sense of community. While Toledo was a source of my American identity, Cleveland and Kent became the place where my Polish and American identities were less in conflict. I was a Polish-born American academic. During my time in Kent, for the first time, I began to identify as a migrant. While the mindset transition was likely more fluid, I seem to associate it with events leading up to my international fieldwork. As I was preparing for my fieldwork in the United Kingdom to study Polish labor migration, I read many scholarly pieces that discussed migrant experiences, push/pull factors for migration, and stories of migrants in which I was able to observe my own life experiences. Finally, the fieldwork process itself helped me to understand that in fact I was a migrant (see Kusek and Smiley 2014).

As I mentioned in the resulting publication (Kusek and Smiley 2014), it was during my fieldwork that I recognized that my own research was heavily influenced by who I was, particularly by my own position in relation to my research subjects—issues previously raised by Chacko (2004), Nagar and Geiger (2007), and Valentine (2002).This positionality includes how I related to those people I was researching in terms of cultural backgrounds and socioeconomic characteristics. As Meriam et al. (2001) note, researchers traditionally are categorized as insiders or outsiders relative to the ethnic or cultural group they are studying, based on whether or not they share characteristics. However, positionality can take more complex forms when past experiences or personal characteristics of the researcher place the researcher in "spaces of betweenness" (Chacko 2004, 54) where their position is fluid and includes both insider/outsider perspectives (see Entrikin 1991; Katz 1994; Nast 1994; Mullings 1999; Chacko 2004).

I experienced a position of "betweenness" when I realized that despite varying cultural, economic, and urban contexts, classifying oneself as either an insider or outsider is not a simple task. This fluidity of insider and outsider status was important to my research projects since it informed how various people interacted with me and helped provide access to certain people and places (Kusek and Smiley 2014).

This simultaneous insider and outsider status was central to my research in London which I conducted between 2011 and 2014. As a Polish native and an immigrant, it was easier for me to assume the position of an insider in the broader group of Polish migrants in Western countries. Yet, by treating London as a research site rather than migrant home, I was an outsider while in the city. To young professional migrants, I was an insider in terms of age and broader life experiences. Like these young professionals I was educated, well traveled, and aspiring for a successful career. When among underprivileged labor migrants, the insider status was more difficult to achieve. On the one hand, we all grew up in Poland during a similar time, characterized first by socialism and later by early capitalism and consumerism. On the other hand, I often felt that my migratory path to the United States was perceived as more advantageous for my career than that of other U.K. migrants.

MARQUETTE, MICHIGAN—PLACE OF LIVING AND ADULT LIFE

I now live in Marquette, Michigan—a place which I am discovering, thanks to my job. I am a faculty member who teaches a variety of human geography and regional geography courses. I am trying to understand how this place will influence my identity and what other places it will lead to.

If judged proportionally in terms of investing my time and effort, my life is unfolding in the United States. I still visit Poland as often as I can, but each visit is characterized by the overwhelming feeling that I am no longer "going back home"—I simply go to visit Poznan and to visit with my family. The feeling of "going" instead of "going back" is strengthened by the fact that Poznan changes year to year, month to month. The city's infrastructure, hubs of cultural and leisurely activities, and even personal networks are being redefined and reshaped on a regular basis. Nevertheless, as a transnational geographer, I am becoming a strong proponent of the theory that "home" is a very fluid term. As I dwell on my life, its past and future, I am beginning to think that that fluidity does not necessarily mean that "home" is sometimes one place, and other times another. Home is maybe rather a place of many meanings simultaneously, or maybe cumulatively. For those of us who are fortunate enough to physically travel, or at least travel through research, pictures, and information, maybe "home" connects to more than one place. As I become comfortable flying often the same routes, at the same airports, driving for hours to and from international hubs, and spending Thanksgiving and Christmas in Toledo, Ohio, summers in Poznan, Halloween in Kent, and the Fourth of July in Marquette, I begin to think that my home is defined by all these places. An inability to continuously connect to all of them would feel to me like having a house with a room I cannot use. Each one of these places has shaped my identity in the past, but I strive to also make all these places a part of my life today.

With this complex definition of home at the core of my being, I try to apply and leverage my "identities" on a daily basis. In my classes I often apply my own personal and professional experiences to teaching geographic concepts. I explain to my students the push and pull factors for my own relocation, I describe the culture shock I experienced when I first moved to the United States, and I use pictures from Poland and England while discussing urban geography concepts. Finally, I try to encourage all to challenge their identities as well. In that sense, the complex identities and places I have navigated help me support the general mission of academia—to teach, to discover, and very often, to challenge.

REFERENCES

Bachelard G. 1994. *The Poetics of Space*. Beacon Press: Boston.
Chacko E. 2004. Positionality and praxis: fieldwork experiences in rural India. *Singapore Journal of Tropical Geography*, 25 (1), 51–63.
Entrikin J. N. 1991. *The Betweenness of Place: Towards a Geography of Modernity*. Baltimore: Johns Hopkins University Press.
Favell A. 2008. Eurostars and Eurocities. Wiley-Blackwell.
Katz C. 1994. Playing the field: questions of fieldwork in geography. *The Professional Geographer*, 46(1): 67–72.
Kusek W. 2014. The Construction and Development of Diasporic Networks by Recent Polish Migrants to London, UK. Doctoral dissertation. Available online: https://etd.ohiolink.edu/ap/10?0::NO:10:P10_ACCESSION_NUM:kent1406713803.
Kusek W. 2015. Transnational Identities and Immigrant Spaces of Polish "Eurostars" in London, UK. *Journal of Cultural Geography* 32, 102–114.
Kusek W. and Radomyski J. 2008. Wrazenia i Opinie z Wielkiej Przygody w Ameryce. In *Spotkania Polsko-Amerykanskie*, Wojciechowski S. and Marszalek-Kawa J. (Eds.).
Kusek W. and Smiley S. 2014. Navigating the city: gender and positionality in cultural geography research. *Journal of Cultural Geography*, Volume 31, Issue 2, 2014. Special Issue: Contemporary Research Strategies in Cultural Geography.
Marcus A. P. 2011. Transnational Rio de Janeiro: (Re)Visiting Geographical Experiences. In *Growing Up Transnational: Identity and Kinship in a Global Era*, 21–35, edited by May Friedman and Silvia Schultermandl. Toronto, Canada: University of Toronto Press.
Mullings B. 1999. Insider or outsider, both or neither: some dilemmas of interviewing in a cross-cultural setting. *Geoforum*, 30 (4), 337–350.
Nagar R., and Geiger S. 2007. Reflexivity and positionality in feminist fieldwork revisited. In Tickell A., Sheppard E., Peck J., and Barnes T., eds. *Politics and practice in economic geography*. London: Sage, 267–278.
Nast H. 1994. Opening remarks on "women in the field" *The Professional Geographer*, 46 (1), 54–66.
Seamon D. 1979. *A Geography of the Lifeworld: Movement, Rest, and Encounter*. New York: St. Martin's Press.
Tuan Yi-Fu. 1991. A View of Geography. *Geographical Review*. 81(1): 99–107.
Valentine G. 2002. People like us: negotiating sameness and difference in the research process. In P. Moss, ed. *Feminist geography in practice*. 116–126. Oxford: Blackwell.

SIX

A Transnational Life Revisited

Stavros T. Constantinou

In this chapter, I revisit the island of Cyprus and specifically my native village of Tsada in the District of Pafos through my life experiences and my memories of the place. This account represents an understanding of place as my lived experience on the national, regional, and local scales and spans a period of more than six decades. I spent the first eighteen years of my life in the village until I graduated from high school in Pafos. Upon graduation from high school, I took the national entrance examinations and having succeeded, I moved to the capital Lefkosia (Nicosia) where I attended the Pedagogical Academy of Cyprus. At this public postsecondary college I trained as an elementary school teacher. At the same time, I was attending Cyprus College for studies in geography. After my college studies, I served my military service for one year which in Cyprus is compulsory. Following my honorary military discharge, I worked as an elementary school teacher for one year in Lemesos (Limassol). In the summer of 1974, when Turkey launched the military invasion of Cyprus, I was called back to military duty as a reservist. By late 1974, I resumed teaching in a two-room school in the village of Dymes, where I taught for two years. I received a Fulbright travel grant and an educational leave of absence from the Ministry of Education of Cyprus, and left my position as an elementary school teacher in 1976 for graduate studies in geography (master's and PhD degrees) in the United States.

Since 1976, I have visited Cyprus almost every year or every other year. Thus, I am a transnational who lives in the United States, still maintaining close ties with my place of birth. My lived experiences have been shaped by the events that transpired in Cyprus during my childhood, adolescence, and early adulthood. Naturally, my experiences of place and space have been altered and shaped by my personal changes during these thirty-nine years and the changes that have taken place in Cyprus. My lived experiences of the spaces in my native village and other places in Cyprus have been altered as I look at space and place through the disciplinary lens of geography, especially as it relates to my research in the areas of migration and ethnicity. My memory sometimes recalls even insignificant past events with laser accuracy, and sometimes becomes very dull. I am surprised by how many of my life experiences I am able to unearth.

Probably few, if any, countries of comparable geographical area the size of Cyprus (3,951 square kilometers) have undergone such profound changes similar to those that I experienced in my lifetime. During the second half of the twentieth century, the island experienced a fierce anti-colonial guerilla campaign (1955–1959), a short period of statehood (1960–1963), inter-ethnic conflict (1963–1974), and outside military invasion (1974) which was followed by "ethnic cleansing" and the separation of the national space into two ethnically homogeneous regions. The inhabitants of the island during these times experienced dramatic events, and this chapter details my personal experiences of these events. Next I will provide a brief account of my childhood as a backdrop of what followed.

I was born in 1951 in the village of Tsada, one of approximately 600 villages and towns in Cyprus. In terms of population, Tsada is one of the larger villages of the District of Pafos. The area of the village is 17.40 square kilometers (6.72 square miles) or 13,011 donums and has the shape of a rectangle. It is located at a distance of eight kilometers (five miles) from the town of Pafos—the seat of government for the District of Pafos. The situation of the village is very important, because most of the traffic to the hinterland of the District of Pafos passes on the outskirts of the village. The topography of Tsada is hilly with very few relatively flat fields. Although the average elevation of Tsada is about 600 meters (2,000 feet), the hilly terrain gives the impression of a rugged mountainous landscape. Driving along the main road from Pafos to Polis was very dangerous, because the old colonial one-lane sinuous road gained 300 m in elevation over a short distance of 3.2 kilometers (2 miles). Because of this very steep slope and the sharp precipice, Cypriots described these as *anefora tis Tsadas* (acclivities of Tsada). The construction of the present two-lane road that replaced the older one-lane road improved the situation significantly.

The residential section of the village is on top of the watershed along the road from Pafos to Polis. An important locational factor was the presence of two water springs, at Rodios on the northern side of the watershed and at Gerolakkos on the southern side of the watershed. The inhabitants of Tsada who lived in the "upper" part of the village depended on Rodios for their water needs, and the people who lived in the "lower" part got their water supplies from Gerolakkos. These springs lost their significance in the 1950s when the authorities built about twenty fountains at different locations in the village. By the 1960s, the fountains were replaced by a system of water pipes that reached all homes.

The views from the hills of Tsada are panoramic. Look to the south from any of the hills of Tsada and you will see the coastal plain and the sea. Look to the north and you will see the Troodos mountains in the distance. The snow-covered highest peak of Mount Olympus is especially visible on clear winter days. In the years since the 1980s, the panoramic views of the sea were particularly attractive to real estate developers who built expensive homes on the hills of Tsada. The houses of the village form a cluster to the east of the main road leading from the town of Pafos to the small town of Polis on the northwestern coast of island. Demographically, Tsada was always a purely Greek village. According to the census of 1946 (the census closest to my year of birth), Tsada had 953 inhabitants (247 households) and they were all Greek Orthodox (Cyprus, Census 1949). Similarly, all villages bordering on Tsada were purely Greek villages. Given that no Moslems lived in Tsada or the surrounding villages, growing up I did not have any interactions with Turkish Cypriots. I remember, however, some of the Turkish Cypriots who served as policemen in the village. What I recall is how well they got along with the Greek Orthodox inhabitants of Tsada. During Christmas, it was customary for each household to slaughter a pig and consume all parts of it during the cold winter months. One time, I was present when a Turkish Cypriot police sergeant who was serving at the village police station, despite the prohibition of consuming pork by Moslems, was the first to taste the freshly slaughtered pig.

The residential part of Tsada was of the nucleated variety. The houses were only separated from each other by a low wall and, thus, everybody knew what was going on in the neighboring households. Our house was in the shape of a long rectangle and was partitioned into five rooms. It was not uncommon to have people and animals (oxen, sheep, and goats) share the same physical space. This arrangement reminded me of a passage of the nativity of Christ in the Bible. An icon of the depiction of this event in the church of the village made it real because my everyday life was not much different. The same is also true of the parable of the sower. In this case, I envision a figure like my father sowing the seeds of wheat in our fields. Like most of the inhabitants of the village, my activity space during my childhood years was the area of the village where I lived with my parents, Theophilos (1924–1999) and Paraskevou (1924–2007), and my five siblings, George, Kostas, Loizos, Andreas, and Marios. A few times during the course of the year, we would travel by foot to some of our fields in neighboring villages for the purpose of tilling the soil or harvesting crops such as wheat or carobs. Before the 1960s, a day trip to the town of Pafos was very rare, because it was done by foot or donkey and would take about one hour and fifteen minutes one way.

By the age of six, I was enrolled at the elementary school of my native village, which I attended from 1957 to 1963. The British colonial administration made attendance of primary school compulsory. Like all students of the elementary school of Tsada, I walked to school and I had to wear a uniform. Classes began in the morning at 8:00 a.m. and after an hour of midday break, they resumed for the afternoon session with dismissal at 4:00 p.m. I recall the major emphasis that the villagers placed on education as the means toward a better life, and the opportunity that they did not have, and as a result, they had to endure a life of deprivation and poverty. In my family, both of my parents, but especially my mother, Paraskevou, were constantly promoting the value of education to me and my siblings. My mother was always lamenting the fact that she did not finish elementary school—she only attended first and second grades—because her parents needed help with tending the family flock of sheep. Because she ascribed such great value to education, she forcefully projected the benefits of education and did not want any of her children to be illiterate. Her mantra was the following Greek proverb: "An illiterate person is like an unhewn piece of wood." My father, Theophilos, fared better, because he finished the village elementary school (six grades) and was able to read and write. Admittedly, the promotion of the value of education worked, because all five of my brothers received a university degree. As for me, I was the first in my family to attend a university and the first in the village to obtain a doctoral degree. Since then five other village natives have obtained doctorates.

The elementary school of Tsada was a three-room building with three teachers and a sizable school yard. The building was new, because it was constructed by the British colonial authorities to replace the one that was destroyed by the earthquake that occurred on September 10, 1953, and caused serious damage in 135 villages located within a radius of 40 kilometers (25 miles) from the city of Pafos, including about 40 casualties (Earthquake 1953). In addition to the elementary school, the earthquake destroyed the historic church of the village that was built in 1908 and contained an old icon of the Blessed Virgin Mary (*Panagia Chryseleousa*), probably dating from about 1540 (Gunnis 1947). The villagers, in an unprecedented effort of solidarity and support, pooled their resources and provided their skills and labor *gratis* and rebuilt the village church next to the location of the one destroyed by the earthquake.

The earthquake also destroyed several houses in the village, including the second story of our stone house. I have a vivid memory of the destruction of our house and the pile of stones that formed in the yard. The psychological impact of an earthquake lasts for a long time. First, the victims have to deal with the physical destruction, which is a lot easier to handle than the fear of the earthquake which lasts for one's lifetime. The colonial administration rebuilt our house, but my fear of another earthquake persisted. Throughout my years in elementary school, I was constantly asking for assurances from my father and mother that another earthquake would not destroy our house and kill us. During my many visits to the village, I had conversations with villagers who experienced the earthquake and recalled details regarding their whereabouts at the time, for an event that happened more than sixty years ago. For a small village, the earthquake became a sort of reference point in their lives.

The colonial administration built special houses to replace those that were destroyed by the earthquake. The villagers called this type of house "barracka," indicating that it was inferior to the house it replaced. For one, the walls of the new houses were constructed of cement blocks instead of stone and the roofs were covered with asbestos. Following the completion of approximately twenty of these houses, the village-scape changed in appearance. Some villagers used these houses as their main residence, others used them for storage. We used ours as a living quarters and as a storage area.

The poor and modest houses of Tsada began to be replaced in the late 1970s and early 1980s. By this time, new technologies combined with shifts in the economy caused major changes in the cultural landscape of Tsada. The construction of new and expensive homes, some with swimming pools, changed the appearance of the village. The tops of rocky hills that were previously undesirable for cultivation have provided the site for expensive homes, because of their ocean views. This is an impressive transformation of the hilly landscape in a relatively short period of time.

During my childhood and teenage years, life in Cyprus was predominantly rural. In the 1950s and 1960s, the villagers cultivated the land around the village with the use of traditional means. A pair of oxen, a yoke, and a wooden plow were the basic means for tilling the thin, and often unproductive soil. An ancient inhabitant of Cyprus would have no difficulty recognizing the Cypriot farmer of my youth. Writing about the first ten years of British colonial rule (1878–1888), Jenness (1962, 149) made this observation: "But the peasants still solidly followed the methods their ancestors had employed in the days of the Pharaohs, using antiquated tools at the cost of back-breaking labor for both man and animal." My grandparents, my parents, and my farmer relatives fit this description.

My parents raised the typical Mediterranean crops like wheat, grapes, carobs, and olives. They also kept a few goats and sheep for milk and cheese, to be used mostly by the family, although sometimes they would sell small quantities. The local cheese is called *halloumi* and recently has been exported to the United States and Europe, especially the United Kingdom. My parents, like most village farmers, also owned a pair of donkeys, the means of transportation around the village. Each villager owned a few fields on which he grew the most basic crops for survival.

Given that the administrative area of Tsada comprises about 3,000 parcels, the average size of a parcel was about 1.43 acres (Department of Lands and Surveys, September 15, 2005). In one exceptional case, a family of eight people owned a single parcel of land that had an area of only 0.2 acres (Leonida 2000).

Not all of the area of the village, however, is subdivided among the farmers of Tsada. The Diocese of Pafos owns an area of 3,848 donums (29.58 percent) extending from the vicinity of the small Monastery of Stavros tis Minthas (Holy Cross), which is approximately 3.2 km (2 miles) from the village, to the southeastern boundaries of the village. This monastery dates from the Middle Ages and the main entrance shows a Gothic influence. This monastery served as the residence of the Metropolitan (Bishop) of Pafos (Gunnis 1947). During his travels in Cyprus, Barsky visited this monastery in 1735 and described the buildings as being in a state of near collapse (Stylianou 1957, 75–76). Furthermore, Barsky mentioned that there was only one monk and one layperson. He also stated that the Bishop of Pafos Joachim was planning to rebuild the monastery, which was done in 1740. In the late 1970s and 1980s, my parents rented fields for cultivation next to the monastery.

The monastery of Stavros tis Minthas celebrates on September 14, the date on which the Greek Orthodox Church commemorates the elevation of the cross. The periodic market (*panegyris*) held here every year on September 14 was small, and participants were mostly from Tsada and the surrounding villages. Typically, the villagers sold and bought farming products and some animals. A bigger periodic market was held at the locality of Agios Konstantinos and Eleni, just outside Tsada, on June 3. On this date, the Greek Orthodox Church commemorates Saints Constantine and Helen, according to the old calendar. An even bigger periodic market was held at Agios Neofytos Monastery. These markets drew participants from a much wider area and offered a bigger variety of merchandise. Periodic markets, usually held at or near monasteries, provided an opportunity for villagers to buy and sell farming equipment, produce, dry fruits, and animals. After all, villages were not as isolated as sometimes we are led to believe. Even at a time when transportation was carried out along poorly developed dirt paths by foot or donkey, villagers created a network of opportunities for trading items that they needed. As children, we looked forward to these markets, because we had the opportunity to buy certain sweets that were on sale at these fairs.

In addition to the area of Tsada owned by the Diocese of Pafos, the Monastery of Agios Neofytos owns an area of 131 donums in the northwestern part around the localities of Agios Antonios and Agios Elias. Farmers from Tsada, who did not own sufficient land to raise crops and support their families, rented plots of land from the Diocese of Pafos and Agios Neofytos Monastery in order to supplement crops raised from individually owned land. Almost every year, my father rented a hilly area (*laoni*) in the locality called "Makrones" from the headman (*mouktaris*) of the village who was subleasing the entire area from the See of Pafos. He alternated between two locations because he wanted the field to remain fallow for one year and thus the soil replenish its nutrients. For the same reason, occasionally, my father would rent property from the abbot of the Agios Neofytos Monastery. The relationships between the villagers and the monasteries on which so many of them depended for their survival were complex and complicated. Whether in rented fields or individually owned, the overwhelming majority of the inhabitants of Tsada had to toil very long hours under undependable climatic conditions to eke out a living. As a family, we spent most of our time in the fields plowing the soil, sowing the seed, weeding, and threshing wheat and other cereals, necessary for our survival. Like all villagers, our relationship with the land was intimate, because our lives could not exist without it. Personally, I am convinced that my interest in geography came about as a result of this close and comfortable relationship with the land during my childhood and adolescence.

The area surrounding the Monastery of Stavros tis Minthas, where I spent most of my early life, has been transformed completely in the last few decades. In 1994, the first 18-hole championship golf course of Cyprus was built on the grounds of the monastery. The fields that I knew as a young boy and teenager have turned into greens, fairways, and water features. A small dam near the monastery serves as a cistern to collect winter rainfall for the watering of the golf course. Water is also carried from a number of wells that were dug in the fields further away from the monastery. There is a clubhouse with a restaurant and practice areas. Several houses have been built on the edge of the golf course, similarly to the golf courses in the United States. Most recently, the golf course has undergone major enhancements and reopened to golfers. Every time I visit my village I jog from the village to the fields that my family used to cultivate beyond the golf course, and I find the change in land use incredible. As a child, I would have never guessed that in my lifetime I would live to see this area put into such a different use from the farming area that it was and that supported so many village families.

Because of the land fragmentation of the village, the farmer spent a significant amount of time traveling by foot or by donkey from field to field. The travel time was even more wasteful for those who owned property in neighboring villages. My family owned such a field in the village of Koili, because my paternal grandmother was from this village. My siblings and I were not very pleased every August that we had to travel to this field to harvest carobs, the main crop on this property.

The villagers measured the value of the land based upon the amount of crops they would harvest, especially wheat, to feed their families. During my childhood and adolescence, farming was a labor-intensive activity, involving both human and animal labor. Farming was mostly of the subsistence variety, although some surpluses supplemented the family income and allowed my parents to buy other necessities of life. As late as the 1950s, mostly human and animal labor was employed in the threshing of wheat. By the late 1950s and 1960s, small implements and other machinery invaded the life of the farmer and eventually displaced human and animal labor.

Both my paternal and maternal grandparents lived in the village and I remember them well. They came from big families and they had big families themselves. My father was one of nine siblings and my mother was one of seven siblings. Having a large family was a common practice during the early decades of the twentieth century. The population of Tsada fluctuated over the years, reaching 943 inhabitants in 1946 (Cyprus. Census 1949). According to the census of 2011, Tsada had 1024 inhabitants.

I have more memories of my maternal grandfather, Loizos Michael Xinaris (1878–1962), and my maternal grandmother, Paradeisa (1884–1961), because our house was very close to theirs and also because we had fields next to each other, and as a young boy I had the responsibility of irrigating a family garden in the field next to theirs. The water issued from a shaft (*lagoumi*) that he dug on the side of a limestone cliff with a shovel and pick in his youth. It would accumulate in a small pond and every other day I would go and release the water to irrigate a little plot of potatoes and tomatoes. Many times he would tell me stories about how difficult his life was and the grinding poverty that people of his generation had to endure.

The most fascinating story that he told me was that he served as a muleteer with the British Army in Salonica Greece during World War I. Although he was very proud of this service and the opportunity to travel to experience another country, he told me that he did it for economic reasons. After he returned to Tsada, the British government awarded him a bronze medal which he had hanging in his closet. The front of this medal depicts the head of King George V and the back depicts a mounted soldier with the date 1914–1918. My grandfather's service number, rank (muleteer), and name of unit (Macedonian Corps) is impressed on the rim of the medal. As stated by Varnava (2014), the Cypriot Mule Corps (1916–1919) was a massively successful program and attracted up to 12,200 men, mostly Greek Orthodox peasants from rural Cyprus. In the case of Tsada, 17.6 percent (55 inhabitants) of the male population of the village in 1911 enlisted in the Cypriot Mule Corps Varnava (2014, 86).

About a century earlier, another ancestor on my maternal side of the family, Christodoulos Kokkinoftas, participated in the Greek War of Independence and took part in the celebrated exodus of Mesolongi on April 10, 1826 (Philippou 1953). He was the grandfather of my maternal grandfather, Loizos. He was a prosperous shepherd in Tsada and after the liberation of Greece he returned to his village. He has been honored in a marble plaque in the Garden of Heroes in Mesolongi. To this day, relatives tell his story with great pride, because of his participation in such a glorious event of the modern history of Greece. The "sacred town" of Mesolongi attracted fighters from many countries to help in the struggle of Greece for freedom against the Turks (the poet Lord Byron is probably the most famous philhellene that died in Mesolongi).

My paternal grandparents, Kostis (1888–1978) and Paraskevou (1906–1972) Kyriacou, and my maternal grandparents, Loizos and Paradeisa Michael Xinaris, were small farmers that worked the land to eke out a living and support their families. My father and all my uncles and aunts who got married and settled in the village continued the same practices. I was one of six siblings, and several of my uncles and aunts also had large families. In the village, I knew all of my relatives well. I got to meet several first cousins many years later, especially those that settled in the city of Lemesos and the capital Lefkosia. In one instance, I met my first cousin Paraskevi, for the first time, in Toronto, Canada, because she emigrated from the village when I was a child.

My childhood memories are those of a difficult, but simple life that was closely linked to a challenging natural environment that offered few choices. Life was indeed hard, but it was hard for everybody, and I was not aware of any opportunities somewhere else. Like everybody else in the village, my siblings and I spent many hours in the fields with our parents helping in many different farming tasks. My parents owned small fields in about seven different locations around the village and two in a neighboring village. The villagers were always worried about undependable and insufficient rainfall for their crops. Cyprus is known to suffer from a serious drought every ten years (Meyer 1962). An example of this was the severe drought (*astohia*) of 1931–1932 that my grandparents and parents often described as devastating. The rainfall for that year was 252 mm (9.92 inches). Crop failure and heavy taxation made life nearly unbearable.

The life of a typical farmer in Tsada was simple and highly predictable. Work in the fields began at the crack of dawn and continued uninterrupted all day, both in the intense heat of the long days of summer and the cold days of winter. Time moved very slowly and the activity space of the villagers covered mostly the administrative area of the village where their properties existed, with only occasional travel to properties in neighboring villages.

COLONIALISM (1878–1960)

By virtue of its vital strategic location, at the crossroads of Europe, Asia, and Africa, Cyprus was occupied by all powers that dominated the eastern Mediterranean. The long list of conquerors consisted of the Phoenicians, Assyrians, Egyptians, Persians, Ptolemies, Romans, Byzantines, Crusaders, Templars, Lusignans, Venetians, Ottomans, and British.

Great Britain acquired Cyprus from the Ottomans in 1878 as part of a lease hold agreement between Queen Victoria and Sultan Abdul Hamid II. The main reasons for the acquisition of Cyprus were the need for a British base near the recurring Russian-Ottoman confrontations and the need to control important trade routes in the eastern Mediterranean, especially those from the Mediterranean Sea to the Red Sea via the Suez Canal and from the Mediterranean Sea to the Persian Gulf overland and beyond to imperial interests in India. From day one of the British colonial rule, Greek Cypriots promoted the cause of *enosis* (union with Greece). Britain rejected these demands, and in 1955, the National Organization of Cypriot Fighters (EOKA) launched an armed anticolonial campaign, which engulfed the entire island. After four years of bloody guerilla warfare, Britain granted independence to Cyprus with the establishment of the Republic of Cyprus on August 16, 1960.

My recollections of this campaign involve the transformation of Tsada into an occupied military village. As I noted earlier, the village is located on the hilly terrain that overlooks the sinuous road that connects the town of Pafos with Polis on the northwestern coast of the island. In an effort to secure the movement of troops to the interior, the British fortified the rural police station of the village that dated from 1905 (Goodwin 1984). In addition, they established numerous sentry posts on top of many of the village hills as lookouts for the main road. The reason behind these British fortifications was to prevent ambushes during the movement of troops along the road. The sight of British troops in full military gear moving in the village was a frightening experience for an elementary school student. I was four years old when the anticolonial campaign started in 1955, and I was in the third grade when it ended in 1959.

While in elementary school, I recall the frequent student anti-British demonstrations. The older students, usually fifth and sixth graders, would unfurl a Greek flag and ask everybody to demonstrate against the British by marching in the school yard or in the streets of the village and toward the police station, which is located at the other end of the village.

Another tactic that the British troops employed during the anticolonial campaign (1955–1959) was the frequent imposition of curfew on the village every time there was an ambush along the Pafos to Polis road. House-to-house searches were common, and innocent villagers would be rounded up in the main square and mistreated in an effort to extract information about fighters who ambushed military convoys that were moving along the road from Pafos to Polis. The topography of Tsada was ideal for this type of activity. General Grivas in his memoirs about the National Organization of Cypriot Fighters (EOKA) lists five incidents of attacks against the British that occurred in Tsada (Grivas 1961).

One of the most celebrated heroes of the anticolonial campaign, Evagoras Pallikarides (1938–1957), was a native of Tsada. His house was in the same neighborhood as my family's house. He attended the elementary school of Tsada and upon graduating his family moved to Pafos where he enrolled in high school at Pafos Gymnasium. My mother told us the story that one time her father (my grandfather) was unloading sacks of wheat on his back in a relative's house, one house from our house. Upon observing my grandfather, teenage Evagoras composed the following four-verse poem for him on the spot: "When a man gets married should not dance/ he should be looking for a sack of wheat on his shoulder." In other words, the concern of a married man should be to provide for his family.

While a high school senior, Evagoras quit school and joined the ranks of the anticolonial fighters. After he was arrested by British security forces for carrying a gun in the forest of Pafos, he was tried and sentenced to death. Despite the many appeals to spare his life, the colonial administration of Sir John Harding executed the nineteen-year-old by hanging in the Central Prison in Nicosia on March 13, 1957. Evagoras was a talented poet, and although he died so young, he wrote many poems, mostly hymns to freedom. I recall the unveiling ceremony of his marble bust in the churchyard at Tsada. Recently, the village authorities turned the Pallikarides house into a museum, and they constructed an open-air amphitheater where the village organizes community events, especially in the summer.

STATEHOOD (1960–1963)

After four years of guerilla warfare, Britain granted independence to Cyprus, and the Republic of Cyprus came into being on August 16, 1960. Spatially, the new state controlled most of the area of the island, with the exception of 256.40 square kilometers (99 square miles), the area of two military bases at Akrotiri and Dekeleia. Furthermore, Greece and Turkey, as guarantor powers, stationed military contingents in Lefkosia. The new bicommunal state was the product of negotiations among Greece, Turkey, and the United Kingdom.

Because of fears about the viability of the newly established state, high unemployment, and the impending passage of the restrictive Commonwealth Immigration Act by the United Kingdom that went into effect on July 1, 1962, Cypriots emigrated in massive numbers during 1960 and 1961 (Diamantides and Constantinou 1989, Constantinou 1990). Emigration touched all towns and villages, and Tsada was no exception. Because of the seriousness of the problem, various organizations started a campaign in an effort to stem the flow of emigrants. I recall a poster that was placed on the wall of one of the coffeehouses of the village with this message: "Compatriot emigrant you will not solve the problem of your life by running away."

INTER-ETHNIC CONFLICT (1963–1974)

After attending Tsada Elementary for six years, I graduated in 1963. I took the entrance examinations, and being successful, I enrolled in Pafos "Gymnasium B." This was the second high school in Pafos, because education was expanding and more students were attending high school. Unlike elementary schools, secondary schools were not compulsory and students had to pay fees. During the last three years, students had to select between either the classical studies or the science track. I chose the science track and I graduated in 1969.

During my six years in high school, I got to meet students from the town of Pafos and other villages. My routine was the same. I would help my parents in the fields on weekends and holidays and commute from Tsada to Pafos to attend school. Getting from Tsada to Pafos was a most difficult task, because there was no regular bus service from the village to Pafos. I would rise early and walk for about ten minutes from my house to the main road from Pafos to Polis to hitchhike a ride to school. This was very hard, especially on frosty winter mornings, and several times I missed classes because I could not get a ride.

An interethnic conflict between Greek and Turkish Cypriots broke out in 1963 with a deadly incident in the capital Lefkosia. In the years that followed, the violence spread in other parts of the island, especially in towns and villages with mixed populations. Violent intercommunal clashes did not involve Tsada directly, because the village had solely Greek inhabitants.

Because of interethnic fighting, Turkish Cypriots, Greek Cypriots, and members of the Armenian community moved from less safe to safer villages or the sectors of towns. According to Patrick (1976, 75), about 25,000 Turkish Cypriots became refugees during the period from December 1963 to August 1964. The immediate result of these movements was the formation of Turkish enclaves in several parts of the island that lay beyond the control of the government. Turkish Cypriots would set up roadblocks to prevent the free movement of people and products. On August 6, 1964, the National Guard carried out an offensive to clear the Kokkina roadblocks and restore to free traffic the coastal highway from Polis to Lefkosia. Turkey interfered with a bombing campaign, and operations halted. Because of the Turkish action, the main road from Polis to Lefkosia along the northern coast remains closed to this day.

On the day of the bombing, I was with my mother watering our garden, and we could hear and see in the distance the Turkish air force jets that were circling in the sky and feared for our lives. These were frightening experiences for a thirteen-year-old. On October 26, 1964, my mother gave birth to my youngest brother Marios. After the delivery, the attending physician voiced the opinion that he will grow up and one day he will go to fight at Kokkina. Immediately, my aunt Eleni countered that she hoped that he is wrong and by the time baby Marios grows up there will be peace and the Cyprus Question will be long solved. To date, the Kokkina enclave continues to block the road from Polis to Lefkosia on the northern coast of Cyprus. I guess the obstetrician was correct, because my brother served his compulsory military service and the Cyprus problem remains unresolved.

I graduated from Pafos Gymnasium B with my high school diploma in June 1969. Given that Cyprus did not have a university at that time—the University of Cyprus was established in 1989—high school graduates who wanted and could afford a university education would attend universities in Greece or other countries. For students like myself, whose parents could not afford to send them overseas, one of the options for post–high school education in Cyprus was the Pedagogical Academy of Cyprus (Teachers Training College), a three-year college for the training of elementary school teachers. Applicants had to pass entrance examinations, written and oral. After passing the exams successfully, I attended the Cyprus Pedagogical Academy from 1969 to 1972. During my studies, I stayed with my uncle Euripides (a younger brother of my father) and his family in one of the new suburbs of Lefkosia called Acropolis. His house was not that far from school, and I was getting there by riding my bicycle. The cityscape of this suburb, like the rest of Lefkosia, has become completely transformed in the years since then. New apartment complexes, detached residential houses, and commercial establishments have filled all of the empty areas that I was driving through while riding my bicycle to school.

While attending the Pedagogical Academy of Cyprus, I was also attending evening geography classes offered at Cyprus College, where I graduated with a BA in geography. The physical geography courses involved several field trips, especially to the picturesque coast of the District of Keryneia in northern Cyprus. The narrow coastal plain extends from the foothills of the Pentadaktylos (Kyrenia) Range to the sea. Although the highest peak, Kyparissovounos, is only 1,024 meters above sea level, it rises like a mountainous wall only about 4.8 kilometers (three miles) from the sea. The celebrated British author Lawrence Durrell (1957, 31) described the range in this manner: "It is *par excellence* the Gothic range, for it is studded with crusader castles pitched on the dizzy spines of the mountains, commanding the roads which run over the saddles between." Although I enjoyed all of the field trips, the most memorable was the one that Professor William Dreghorn led east of the Keryneia castle along the beach. During the period 1969 to 1972, I also got to know the beaches around the villages of Panagra, Vasileia, and Orga where I used to go swimming with my uncle Euripides and his family.

All of the District of Keryneia, along with parts of the Districts of Ammochostos, Larnaka, and Lefkosia, has been occupied by Turkey since the Turkish military invasion of Cyprus on July 20, 1974. Although since 2003 people are allowed to cross the dividing line and visit the occupied area, I have not gone to see the Turkish-occupied northern part of the island. Forty-one years later, Cyprus remains divided along the Attila line established by the Turkish military in 1974.

After graduation from the Pedagogical Academy in 1972, I served for one year in the National Guard. Military service in Cyprus is compulsory for eighteen-year-old men. Usually, high school graduates join the military after graduation. Because of a three-year deferral to finish my studies at the Pedagogical Academy of Cyprus, I joined the military when I was 21 years old. During my service, I was stationed in Pafos, Diorios, and Lefkosia. I spent most of my military service (nine months) at a camp near the village of Diorios. During my time there, I got to visit the nearby villages of Kormakitis, Myrtou, and Agia Eirini. I also got to know the area of Agios Kassianos, one of the quarters of the city of Lefkosia, within the walls where I served the last month of my military service. The capital of Cyprus, Lefkosia, is one of the few divided capital cities in the world. The Green Line, drawn in 1964, runs through the city and separates the Greek from the Turkish sector. This line came about as a result of the outbreak of intercommunal violence between the Greek and Turkish communities of the island in 1963. Following adoption of Resolution 186 by the UN Security Council in 1964, a peacekeeping force (UNFICYP) was put in place to keep the peace. Canadian troops patrolled the Green Line in the area where I was stationed.

After my honorable discharge from the National Guard in 1973, I started my first teaching job at an elementary school, teaching first grade in the suburb of Agios Ioannis in the town of Lemesos (Limassol). During the academic year 1973–1974, I had the opportunity to meet my aunt Elisavet (Elizabeth) and her six grown children who moved to Lemesos in the 1960s in search of jobs that were not available in Pafos. I got to know all of my first cousins and their children very well. Because of them, my stay in Lemesos was made easier, and several of these relationships lasted long after I was gone from Cyprus. Kinship is an important dimension of Cypriot family life.

In 1974 I moved from Lemesos to Tsada for the summer vacation, because I did not know where my next teaching assignment was going to be. In Cyprus, an ad hoc committee at the Ministry of Education publishes teaching assignments by August. While I was making plans on ways to spend my summer holidays, two extraordinary events created chaotic conditions in Cyprus: the military coup in 1974 that overthrew the government of Archbishop Makarios III following the Turkish military invasion. I was on my way to the beach in Pafos, when I heard the news on the radio that Greek officers who were staffing the National Guard staged a coup d'état against the government of Archbishop Makarios III. In the uncertainty and confusion that followed, I returned home where I kept a wait-and-see attitude. Makarios survived the coup and fled to his native Pafos from where the British flew him to safety in London.

TURKISH MILITARY INVASION (1974–PRESENT)

Using the coup as a pretext, Turkey launched a full-scale military invasion of Cyprus on July 20, 1974. The first Turkish troops landed on the coast of Keryneia, and additional troops parachuted into the Mesaoria Plain between the Turkish sector of Lefkosia and the Pentadaktylos Range. The first goal of the invasion was to establish a bridgehead extending from the coast of Keryneia to Lefkosia. Once this was established, the Turkish forces kept pushing outwards to expand the area under their control. At the same time, the Turkish air force was bombing towns and villages ahead of the advancing troops. The major objective was to spread fear among the civilian population so that they abandon their homes and properties. After the first phase of the invasion, inconclusive negotiations in Geneva led to the launching of the second phase of advancement by the Turkish army which captured about 38 percent of the area of Cyprus. By the time hostilities ceased, Cyprus was divided along a line that extended from the area around Lefka in the west to the town of Ammochostos in the east. After the ceasefire, Turkish Cypriots from towns and villages in the area still under the control of the Republic of Cyprus were transferred to the northern part of the island under the control of the Turkish army. Thus, ethnic cleansing created two ethnically homogeneous regions. In November 1983, the occupied part of Cyprus proclaimed itself as the Turkish Republic of Northern Cyprus (TRNC). To date, only Turkey has extended diplomatic recognition to this state. I was an eyewitness in Pafos to the convoy of trucks loaded with people and their personal belongings heading for the northern part of the island to start new lives in houses and properties that were evacuated by the fleeing Greeks.

The cost of the invasion was inestimable in terms of loss of life and property. Approximately two hundred thousand people became refugees in their own country, and tent camps were established to provide shelter for those who did not have any relatives to take them in. Families were divided and scattered in different parts of the island. The economy of the island reached a point of near complete collapse. The shock was immense to both the demography and the economy of the island.

By September, I received my new appointment to teach at the elementary school of Dhymes, a small mountainous village (elevation about 1,000 meters) in the Troodos Mountains. According to the census of 1960, the total population of Dhymes was 373. The village had a two-room school, with the first three grades in one room and the upper three grades in the other room. Being the junior teacher, I was teaching the first, second, and third grades, and my colleague was teaching the fourth, fifth, and sixth grades. I worked in Dhymes for two years, 1974 through 1976.

During my tenure at Dymes, I set up a chapter of the Society of the Friends of Trees. The main project of this chapter was to organize the villagers and plant about 200 trees on the newly asphalted road leading from Dymes to Kyperounta. The idea was to plant trees on the outer side of the road to prevent erosion and provide aesthetic value. Raising trees in a dry environment like Cyprus requires protection of the small seedlings from the villagers' goats and watering during the dry summer season. The carpenter of the village used his equipment to build protective cages for the young seedlings. The community secured equipment for the watering of the plants until they topped the cages. I have visited many times since I left and the trees have done extremely well.

In August 1976, upon receiving a Fulbright grant, I left for the United States to pursue graduate work at Western Kentucky University in Bowling Green, Kentucky. I graduated with a master's in geography in 1978 and an Associate of Science degree in Cartography. Upon graduation, I moved to Kent State University, from where I received my doctorate in Geography in 1982.

After finishing my PhD degree, I got a job as a university professor and lived the life of a transnational, in the sense that I maintained close contacts with my family in Cyprus through frequent visits, telephone conversations, and more recently through Skype and Viber. Furthermore, I carried out research on issues of the population geography of Cyprus, which provided a professional connection with academic institutions and research centers on the island. To date, I have presented four papers at international conferences held in Cyprus in 1986, 1993, 1996, and 2008, which have been published in the proceedings of these conferences. One of these papers deals with emigration from Cyprus (Constantinou 1990) and two deal with the evolution of population centers (Constantinou 2001, 2012). The fourth presentation focused on Greek American networks (Constantinou 1996).

In addition to the presentations and publications, these conferences provided ample opportunities for travel around the island to observe economic, social, and political changes that have been under way in Cyprus since my departure in August 1976. The cultural landscape of the island has been transformed completely with an infrastructure that resembles that of advanced Western countries. New four-lane highways connect the main towns, two airports exist at Larnaka and Pafos, and the modern port of Lemesos handles large amounts of cargo and passenger traffic. The airport at Larnaka was built in record time, following the Turkish invasion, to provide an air connection for the island with the outside world. The main airport at Lefkosia became inoperable after Turkish advances close to the airport.

During the 39 years since I left, Cyprus has been transformed from a predominantly rural society to a service-oriented economy, especially tourism. As I travel around the island, I have a feeling that I am traveling somewhere in the US. Probably the major difference is the road signs and the extreme heat of the summer. I recall one time that I was driving from Pafos to Lefkosia with my young son in the car and got behind a McDonald's truck heading to Lefkosia. I said to my son, this is the ultimate in globalization. Needless to say, when I left Cyprus in 1976 I had never heard of McDonald's or any of the other American fast-food restaurants. Now, in every city of Cyprus, including my hometown of Pafos (population 32,892 in 2011), one can find just about every one of the American fast-food chain restaurants. Today's young people of Tsada and Cyprus in general, in order to see a donkey, will need to visit the zoo. They have been replaced with German luxury cars. Walking the streets of Tsada today, one can see people walking their dogs, instead of goats, donkeys, oxen, and sheep. When the society was predominantly rural, hunters and shepherds were the most likely people to own a dog.

During my frequent visits to Cyprus I participate in many planned family events or events that were scheduled when I was visiting. Here is a small anthology of the most important transnational events. I have served as a godfather for one of my nephews at a baptismal ceremony in Crysorrogiatissa Monastery. I participated in a few weddings of the children of my cousins. I participated in many trips to archaeological sites and monasteries with family members and relatives. In 2000, I traveled with my mother and my fourteen-year-old son Theophilos to the Holy Land in Israel and Egypt. In 2004, I traveled with my mother to Turkey to visit Hagia Sophia in Istanbul and the ruins of Christian churches in Asia Minor. I traveled to Cyprus with my son Theophilos for the funeral of my father in 1999 and with my daughter Alexandra for the funeral of my mother in 2005.

Every time I visit Cyprus and Tsada, I feel that I revisit my "old home," which provided sustenance and security for the first twenty-four years of my life. The village of Tsada is where my roots are located, where my parents, grandparents, and other ancestors and relatives that I knew are buried. As I often tell my friends and relatives, most of the villagers I knew have by now transferred to the village cemetery. The physical space of Tsada and Cyprus has changed significantly, but so have the people that inhabited the space and place I knew. I often wonder who changed the most, me or my "old home." My "new home" is in the United States, where I have spent the last thirty-nine years of my life.

This is where I created and raised my family. In this sense, I inhabit my "new home" and revisit my "old home" periodically and relive my childhood and youth with friends and relatives.

REFERENCES

Constantinou, Stavros T. 1990. "Economic Factors and Political Upheavals as Determinants of International Migration: The Case of Cyprus." In *Praktika tou Protou Diethnous Symposiou Kypriakis Metanasteuses: Hestorike Kai Koinoniologike Theorese Leukosia, 29-31 Augoustou 1986*, edited by Cyprus Research Centre, 139–172. Leukosia: Imprinta Ltd.

Constantinou, Stavros T. 1996. "Greek American Networks." In *Les Resaus des Diasporas. The Networks of Diasporas*, edited by Georges Prévélakis. Nicosia: Cyprus Research Centre (KYKEM).

Constantinou, Stavros T. 2001. "Evolution of the Population Centers of Cyprus, 1878–1960." In *Praktika tou Tritou Diethnous Kyprologikou Synedriou (Lefkosia, 16–20 April 1996) [Proceedings of the Third International Congress of Cypriot Studies (Lefkosia, 16–20 April 1996)]*, edited by Ioannis Theocharides, 187–222. Nicosia: Society of Cypriot Studies.

Constantinou, Stavros T. 2012. "Population geography of Cyprus: a comparative spatial analysis of the censuses of 1881 and 1960." In *Praktika tou Tetartou Diethnous Kyprologikou Syndedriou (Lefkosia, April 29–May 3, 2008)*, edited by Eleftherios Antoniou, 817–826. Lefkosia: Society of Cypriot Studies.

Cyprus. Census Office. 1949. *Census of population and agriculture 1946: Tables*. London: O.F. Hodgson.

Department of Lands and Surveys. 15 September 2005. Email to author. Lefkosia.

Diamantides, Nicholas D., and Stavros T. Constantinou. 1989. "Modeling the macrodynamics of international migration: determinants of emigration from Cyprus, 1946 - 85." *Environment and Planning A* 21: 927–950.

Durrell, Lawrence. 1957. *Bitter lemons*. New York: Dutton.

Earthquake, in Western Cyprus. 1953. "Earthquake in Western Cyprus." *Nature* 172: 524.

Goodwin, Jack C. 1984. *Tsadha*. 4 vols. Vol. 2, *An Historical Toponymy of Cyprus*. [Nicosia]: MAM.

Grivas, Georgios. 1961. *Apomnemoneumata Agonos E.O.K.A. 1955-1959 [Memoirs of Struggle of E.O.K.A. 1955–1959]*. Athens.

Gunnis, Rupert. 1947. *Historic Cyprus : a guide to its towns and villages, monasteries, and castles*. London: Methuen.

Jenness, Diamond. 1962. *The Economics of Cyprus: A Survey to 1914*. Montreal: McGill University Press.

Leonida, Panikkos. 2000. *Me Gymna Podia: Odoiporiko Mias Zoes*. Lefkosia: Author.

Meyer, Albert Julius. 1962. *The economy of Cyprus, Harvard Middle Eastern studies, 6*. Cambridge: Harvard University Press.

Patrick, Richard Arthur. 1976. *Political geography and the Cyprus conflict, 1963–1971*. Waterloo, Ont.: Dept. of Geography, Faculty of Environmental Studies, University of Waterloo.

Philippou, Loizos. 1953. *Kyprioi Agonistai [Cypriot Fighters]*. Lefkosia: Holy Archdiocese of Cyprus.

Stylianou, Andreas. 1957. "Travels of Barsky in Cyprus." *Kypriakai Spoudai* 21: 1–139.

Varnava, Andrekos. 2014. "Recruitment and Volunteerism for the Cypriot Mule Corps, 1916–1919. Pushed or Pulled?" *Itinerario* 38(03): 79–101. doi: doi:10.1017/S0165115314000540.

SEVEN

The Urban Geography of My Canadian Identity

Heather A. Smith

I am Canadian—born and raised on the edge of the Scarborough Bluffs in Toronto, Canada. From the front porch of my parents' house you can see the great expanse of Lake Ontario and on very clear days the border with another country. On the clearest of days, often in the midwinter, the horizon sharpens and you can just make out the shape of Rochester, New York's shoreline with what I always assumed were the stacks of waterfront factories billowing smoke high into the cold, crisp air. As a child, that far shore fascinated me, and I used to dream of the Great Lake at my doorstep narrowing its girth so that I could walk across the ice and see what there was to see.

Growing up in Toronto in the late 1970s and early 1980s, I saw the United States, and in particular its cities, as dirty, violent, and angry places. This image was produced and reproduced by the ready, if not overwhelming, presence of U.S. television shows streaming into Canadian living rooms. *Starsky and Hutch* (1975–1979); *Cagney and Lacey* (1981–1988), and *Hill Street Blues* (1981-1987) were weeknight staples, followed by Buffalo, New York's WKBW's late-night news and its anchor Irv Weinstein beginning the broadcast by asking the audience, "It's 11 o'clock, do you know where your children are?" Even *Saturday Night Live*'s opening shots were of the nighttime city, slick with rain—beckoning viewers on to something illicit.

This youthful perception of American cities stood in stark contrast to my lived experience in the Canadian city. From about the time I was twelve years old (1978), I was riding the buses, streetcars, and subways of the Toronto Transit Commission (TTC) to make my way to sports practice, Saturday morning activities at the Royal Ontario Museum (ROM), art classes at the Art Gallery of Ontario (AGO), or after-school theatre rehearsal. The city for me was an exciting place, full of freedom and wonder. I loved taking the 501 streetcar along the span of Queen Street East as it clattered through the Beach neighborhood where I went to school, ran south of the very working-class neighborhood where my grandparents lived, rumbled past the warehouses that serviced the factories and container ships docked at the Port. As the streetcar crossed the bridge spanning the Don River and Don Valley Expressway, it entered the city core and ran along the southern edge of Cabbagetown, then in its very earliest stages of gentrification. Boarded up, graffiti-decorated houses were sandwiched between others with newly planted gardens and refurbished wrought iron stair railings. The stops at Sherbourne and Jarvis Streets introduced rooming houses, pawn shops, bars, and the high-rise, high-density public housing community of Moss Park to the landscape. And then, there was Yonge Street—at that time the longest street in the world—and its promise of the Eaton Centre shopping mall and its brand-new *18-screen Cineplex* movie theatre. While I had certainly been taught to be street-smart and always had a dime in my pocket with which to use a pay phone if needed, fear of the city and the people I encountered in it was not a reality for me. My reality was partial—and privileged.

It was not until college, and a very unexpected move to the United States, that I came to realize that my understanding of both the Canadian city in which I lived and the American city I perceived were mythical and far, far too simplistic.

THE TAR HEEL STATE

My mother was a high school English and Art teacher, my father a Staff Superintendent with the Metropolitan Toronto Police Force (note above the list of regularly watched TV shows). When I thought about attending college, law seemed a wise path and a university within commuting distance to Toronto a frugal choice. Fate and geography intervened.

In 1985, I was presented with the extraordinary opportunity to attend the University of North Carolina at Chapel Hill through the generosity of the John Motley Morehead Scholarship. Modelled after the Rhodes, the Morehead-Cain (as it is now called) provides full-ride support to attend Chapel Hill for four years of undergraduate study. In the mid-1980s, the program was beginning to expand its global reach beyond the United States and Great Britain. I was among the first students from Canada to be offered the scholarship. A distinctive feature of the program was—and remains—its support of summer experiences, which in my era immersed scholars in leadership development, public safety, private enterprise, and international travel. Through the Morehead, I participated in Outward Bound in the Blue Ridge Mountains of North Carolina and Tennessee, worked at the headquarters office of Pan American Airlines in New York City and traveled to Buenos Aires, Argentina, and Bridgetown, Barbados, and spent a summer exploring the cities of Western Europe. And, most significantly in terms of my evolution as an urban social geographer, I interned for the Los Angeles Police Department, where for much of my time I was based out of the Southeast Division serving Watts. On one of the first days of my internship, I was taken to the scene of a gang-related shooting in which a very, very young man lay fatally wounded at the side of the road after boarding a city bus on which rival gang members were riding. Within the first week, I was helping a police sergeant lift a dead body out of a swimming pool and the coroner inspect the corpse and take internal temperatures to determine time of death. This was followed by stints with the drug squad and the child abuse and homicide units. I was nineteen years old.

To say that my experiences at Carolina were transformative would be an understatement. I arrived for my freshman year a young, naïve, and fiercely proud Canadian with no understanding of the dynamics of race and place in the U.S. South, or anywhere else for that matter. I was bewildered by the socio-spatial divisions I noticed on campus and in the surrounding community. It wasn't that such divisions didn't exist in Toronto (they certainly did), but rather that the scale of the city, and the youthful and privileged nature of my experience in it, rendered those divisions "invisible" to me. In Chapel Hill, my dormitory and many of those immediately around it on North Campus were almost exclusively white. Whites, I observed, were residents of the red brick and ivy-covered dormitories closest to the historic center of campus; most blacks, it seemed, lived on South Campus, a distance from classrooms and in high-density, breeze-blocking residential towers that were more racially diverse. The sororities and fraternities dominating the college's social scene were seemingly all white. Students were staging protests in support of a freestanding Black Cultural and History Center distinctive from the already existing Student Union. These observations framed my first awareness of what segregation looked like, and the more questions I asked the more complicated things became. Why would some black students want their own cultural center or student union and prefer to live on South Campus? What criteria was the university using to assign particular students to particular dorms? Why were sororities and fraternities all white? What was the Greek system all about, anyway? And, if there were sororities and fraternities for black and other minority students, where were their ivy-clad houses with rocking-chair front porches and members-only dining rooms?

Other experiences were even more unsettling. Within days of my arrival in Chapel Hill, a young woman and graduate student at UNC was kidnapped at knifepoint from a parking lot fronting the main, and very busy, street of the town. She was subsequently murdered and her body disposed of in an old barrel. I called home and relayed the story to my parents, mentioning that Labor Day was looming and that the university dorms would likely be empty. Within minutes of that phone call ending, my parents had booked my ticket home for the first long weekend of my college career. I remember that flight well. As the plane crossed over Lake Ontario on its approach to Pearson International airport, it passed alongside Toronto's Central Business District and the iconic CN Tower. I was overwhelmed with joy and relief at being home. That autumn day was so clear that looking out the plane window, I could see the sandy white cliff face of the Scarborough Bluffs and atop it the neighborhood where I lived.

TORONTO THE GOOD

By the mid-1980s, Toronto was on its way to becoming one of the most multi-cultural cities in the world. Between 1971 and 1986, the percentage of Torontonians who identified as European dropped from 96 percent to 82 percent. Arabic and West Asian percentages rose from 0.1 percent to 1.0 percent; South Asian from 0.2 percent to 3.5 percent; East Asian from 2.3 percent to 6.6 percent; African from 0.5 percent to 3.5 percent, and Caribbean from 0.6 percent to 1.2 percent (Orstein 2006).

While these proportional counts are nowhere near where they are today in Toronto, the changing ethno-racial geography of the city was the context in which I had been raised. My grandparents lived on Hiawatha Road (my grandmother and mother were both born and raised in the very same house), which ran directly south from Gerrard Street—the heart of a vibrant and rapidly growing South Asian community. The family of one of my early high school friends owned a restaurant a block away from the Bollywood movie theatre and beside the colorful row of sari shops that fronted the Indian Bazaar. Homework sessions in her living room were always accompanied by plates of Punjabi sweets from the restaurant below. I had my very first tandoori in one of the many Indian restaurants lining Gerrard Street. Across the road was a grocery store owned by a Chinese Canadian family who were long-standing friends of my grandmother and had employed my uncle as a delivery boy in the early 1960s. Their children were also friends of my youth. All three of us were in the band at our high school, and for a time a routine developed where woodwind practice would take place at my grandparents' house. This was followed by a brief walk around the corner to dinner. In the kitchen of the flat above the grocery store, I learned how to use chopsticks and developed an enduring love of bok choi. Other friends and schoolmates had families that had migrated from Jamaica, Greece, and England. In each of their homes, markers of ethnicity and migration were present; accents, grandparents who spoke little English, styles of dress, icons of faith, traditions of food, and cultural celebrations that were very different than those practiced in my own family.

Given my name, it should come as no surprise that the dominant thread of my cultural heritage comes from Scotland. My maternal grandfather, William Carnegie Graham, immigrated to Canada when he was a very young boy, with his newly widowed mother and one of his sisters. Another sister was left behind to join them later when further funds for passage could be secured. My grandfather told me that my great-grandfather had been killed in WWI. The story he told my mother and her sisters was that he had been run over by a beer wagon in the cobbled streets of Glasgow. Although having spent all but his very early childhood in Canada, my grandfather was Scottish to the bone. When not working at Amalgamated Electric, he was a drummer in the Toronto Scottish Pipe Band. An enduring memory is watching the arc of his drumsticks as he tossed them high into the air in between beats timed to the steps of his march. At his cottage on Sturgeon Lake in Ontario, he erected a flagpole upon which three flags were always raised when the family was in residence—the Canadian, the flag of the Royal Canadian Air Force (for whom he had served as a bomber navigator in WWII), and the Union Jack. When I was sixteen years old, as he had done for my grandmother, mother, aunts, and cousin before me, he bought me my first kilt—the navy blue, white, and green striped tartan of Graham of Montrose. He also tried in vain to teach me how to play the bagpipes. I was never able to get past the chanter. My grandmother (whose parents were born in what is now Newfoundland, Canada) was a nursing assistant at the Toronto General Hospital where many of her nursing, orderly, and janitorial colleagues were recent immigrants. These individuals had often left families behind and were on their own for Canadian holidays. The dining room table at my grandmother's house (and occasionally our own) expanded to include them. Over Thanksgiving turkey, Christmas morning steak and eggs, Easter ham, New Year's roast beef (all bought from Brown Brothers Meats in Toronto's St. Lawrence Market), we heard stories of other countries and cultures, tales of migration and of the families and loved ones that had been left behind. So, the contrast between the multi-cultural Canadian city in which I had been raised and the biracial southern American college town in which I was attending university was stark and baffling.

DISCOVERING GEOGRAPHY, DISCOVERING SELF

Back in Chapel Hill, I was propelled towards classes that might help me make some sense of the dynamics of race and place that I was encountering in America. This led me to Saunders Hall, the Department of Geography, and the classrooms of Drs. April Veness and John Florin. The geographers at UNC Chapel Hill took good care of me—guiding me in directions that would fundamentally transform my life and set the foundation for my career.

Dr. Veness encouraged me to write about my experiences in Los Angeles as grounded examples of the theories we were learning about in class—homelessness as an outcome of structure *and* agency, the territoriality of gang violence, invasion and succession as an increasing number of Korean descent were setting up home and businesses in traditionally African American neighborhoods, the social construction of race and the multiple forms of racism, the presence and power of community amidst the brutality of poverty and intergenerational disadvantage. It was through geography that I was able to puzzle out and better understand the causes and implications of what I had witnessed and experienced during my time with the Los Angeles Police Department, and come to appreciate the distinctive post-modern nature of the city in which they had occurred.

One of the lessons of my experience in Los Angeles was that impoverished neighborhoods in large American cities were very different from the stereotypical and negative images I had developed in my youth. The defining residential landscape of the South Los Angeles area in which I was working was not high-rise, high-density, and derelict public housing blocks, but rather single family homes with gated font yards edged by sidewalks (that people actually used) with the occasional palm tree. It was familiar and comfortable to me—in some ways (take away the palm trees, add some snow) it wasn't entirely dissimilar to Scarborough and some of the other outer Toronto suburbs. How could somewhere that looked so similar be so very different?

Certainly the nature of my internship exposed me to extreme hardship, violence, and brutality, but it also, and more powerfully, provided the opportunity to meet with residents and, if I was really lucky, to hear a little something about their daily lives. Often conversation would drift into territory about their perceptions of their neighborhood and the vast and sprawling city in which it was embedded. One of the many things that struck me about these conversations was that their experience and perception of the city in which they lived was like mine for the city in which I usually lived—limited and partial. When I told people where I was living while in LA (Santa Monica) they would often say "I've never been there—what is it like?" For those who did not own or have access to a car, movement through the city was dictated by pedestrian or public transit routes. And, like my rides on the 501 streetcar, they too were moving through space, observing the passing city through a glass window, often just getting from one place of familiarity to another.

My interest in geography deepened further when Dr. Florin invited me to attend a departmental guest lecture by a visiting geographer from the University of British Columbia in Vancouver, Canada. I remember vividly sitting on the classroom floor, surrounded by a standing room only crowd, transfixed as David Ley talked about his work on gentrification in Canadian cities. A passion for all things urban and geographic was solidified, but Chapel Hill, at that time, didn't have a full-time urbanist on staff. Dr. Florin suggested that I spend my junior year abroad at Bristol University in the United Kingdom. Between 1987 and 1988 I spent a magical academic year auditing Peter Haggett's spatial diffusion class and learning the foundations of urban economic geography from Nigel Thrift, for whom I wrote my very first research papers on London as a world city, circuits of capital, and gentrification. I still have those papers—written in longhand.

While it was experiences in the classrooms of Chapel Hill and Bristol that led to my becoming a geographer, it was my experiences in Los Angeles that led me to be an urban social one. And one that was especially keen to use my new disciplinary knowledge to critically engage with the cities of home. Gentrification and its unfolding in the Toronto neighborhood in which I had gone to high school became the topic of my undergraduate honors thesis; the causes and consequences of female-headed lone parent families housed in poorly serviced suburban public housing complexes in Toronto the focus of my master's. My doctorate, supervised by the extraordinary David Ley, explored socio-spatial polarization in Vancouver's Gastown and Downtown Eastside, and our collaborative work through the Vancouver Center of Excellence for Research on Immigration and Integration in the Metropolis (RIIM) focused on the intersections between concentrated immigrant settlement and extreme poverty in Canada's three largest cities: Toronto, Montreal, and Vancouver.

SEA, SKY, AND SKID ROW

Vancouver. I can barely write that word without experiencing a wave of nostalgia and homesickness. It is not that I do not pine for Toronto, I do. It was in Vancouver though that I grew into adulthood, established enduring personal and professional relationships, and made critical decisions about how I wanted the balance of my working and home life to take shape. It was also in Vancouver that I learned how to be the kind of academic whose scholarly work had impact in the realms of policy and people's everyday lives.

Newly married (to a London-born Englishman I met while at Chapel Hill), I arrived in Vancouver in the fall of 1992 to begin my PhD in geography at the University of British Columbia. My husband had secured a job with a publishing company and would be setting up and staffing a regional distribution office for the West Coast and western Canadian provinces. It was an exciting time. I had been granted a Social Sciences and Humanities Research Council (SSHRC) doctoral fellowship so financially we were in an advantaged position compared to my fellow graduate students. We were able to rent a one-bedroom apartment in a subdivided house on the border of Kitsilano and West Point Grey. For the first few years of my PhD study I rode my bicycle to my office on campus. It was a roughly six-kilometer ride along the edge of the Burrard Inlet through Jericho Park, up the hillside road of NW Marine Drive (passing by the stunning architecture of one of the city's most exclusive neighborhoods), past the Museum of Anthropology and Japanese Memorial Gardens and into the department of geography at 1984 West Mall. With sea and mountain views all along the route, it was glorious. I knew I never wanted to leave.

Even then, Vancouver was an astoundingly expensive place to live. Friends in the publishing industry and non-profit or service sectors were increasingly challenged in their efforts to purchase or even rent housing that was centrally located to their places of work. Several of my fellow students lived in damp and cramped basement apartments where the lack of sunlight was at times a health issue. Others were lucky enough to secure spots in rarely available student housing on campus. We too were relieved when, after becoming pregnant with our first child, we successfully applied for one of the rare family units to come open.

Affordability in the city was also an issue I was exploring in my dissertation research. Gentrification in its many guises had by that time transformed much of Vancouver's central and Westside neighborhoods and placed them out of reach for those with anything less than a globally competitive professional, cultural, or entrepreneurial income (Mills 1988, 1991; Ley 1988). In Gastown and the Downtown Eastside located on the water's edge adjacent to the city's expanding Central Business District, I was interviewing young urban professionals who spoke about their conflicted decision to invest in property, and live, in an area of the city known primarily for its extreme poverty, open drug use, and epidemic levels of HIV, AIDS, TB, and hepatitis. For many, this was the only area left in the city in which they felt they could afford to live, and they were angry about it. I was also interviewing long-term residents of Gastown and the Downtown Eastside, some of whom lived in the areas' many Single Resident Occupancy (SRO) hotels. They were angry too. From their perspective, the arrival of young professionals made their already very precarious housing even more so. Not only did they feel the threat of eviction given the increasing rate with which buildings were being bought and converted into condominiums for the luxury market, they also spoke of declining conditions in their buildings. The speculation was that landlords, anticipating a future sale, had stopped investing in maintenance and upkeep—classic rent gap. This view was supported by interviews with police officers and housing advocates, both of whom were engaged in the very hard work of helping the community navigate through the conflicting pressures of simultaneous upgrading and downgrading. Ultimately, the lack of affordable housing, the high cost of living, and limited professional mobility in our respective fields led my growing family to look elsewhere for work and a place to call home. I was heartbroken at having to leave.

A RETURN TO CAROLINA

While I returned to Canada for graduate training, it was back to the United States, and we settled in North Carolina, built careers, and raised a family. As I approached the end of my doctoral degree, there were very few Canadian academic positions open or likely to open in my field. When I was fortunate enough to be interviewed for the one Canadian position that did come open, it was clear from the moment I was picked up at the airport that the position was intended for the spouse of someone already on faculty. So, like many before me, I followed the well-worn path south looking for my first tenure-track position in the United States (Lucas 2006; Hardwick and Smith 2012; Hardwick 2014; Swanson 2014). My husband and I talked at length about what we would need to make such a move work. We restricted our search to the coasts and the "new" rather than the "old" or Deep South. Furthermore, we knew we wanted to be either in, or within easy driving distance to, a large city with an airport that had direct flights back to our parents and siblings in Toronto and London. Of the very few positions that met our criteria, UNC Charlotte was looking for an urban geographer, and the match between my interests and their needs seemed perfect. So too, after my interview, did the department's laid-back feel, collegial atmosphere, and interest in building a PhD program. They offered, I accepted.

In July 1999, we put everything we owned (very little) into a U-Haul van, packed up our three-year-old daughter and her grandparents and drove 1,220 kilometers from Toronto, Ontario, to Charlotte, North Carolina. As we crossed the U.S./Canadian border my husband swears he heard the sound of my "brain draining." The U-Haul broke down just south of Erie, Pennsylvania, in the midst of an almighty thunderstorm. We spent a sleepless night in the only hotel room we could find, contemplating our future and second-guessing our choice.

On the day we arrived in Charlotte, we purchased our first house and our first new car. The cost of living in Charlotte was far less than in

Toronto or Vancouver. It needed to be, since we were now reduced to one income.

CROSSING BORDERS AND COMPLICATED VISAS

I had come into the United States on a non-immigrant TN-1 visa, the NAFTA Professional visa that allows citizens of the United States, Canada, and Mexico who have a job offer in hand, and the degree qualifications to support that offer (usually at least, a Bachelor's degree), entrance into and the right to work in the United States on a temporary basis. My visa was for an initial but renewable one-year period. My British-born husband entered the country as my dependent. After more than a decade working in the publishing industry, and several years supporting me while I completed my graduate degrees, he happily stepped into the role of primary household/family manager and caregiver for our daughter.

Visas and documentation status loom large for immigrants—and in the post–9/11 era, perhaps especially so for those who migrate to the United States (Rossiter 2014). The idea of having my family's livelihood tied to a visa that was temporally fixed and by no means guaranteed was daunting; its reality, not risk free. I was delighted and not a little relieved when my university began the process of switching me over to an H1B. This visa, although it required my putting together a dossier akin to a tenure package (including external references commenting on the degree of my professional impact), provided a three-year period for renewability and a path towards permanent residency and the securing of a green card. We were even more relieved when my husband was invited to meet with a national book distributor headquartered in Charlotte. The meeting led to a job offer and a U.S. work visa that was entirely separate from mine, which turned out to be a very good thing. My husband's company helped with green card applications, and when a change of staffing at the university translated into my H1B renewal paperwork not being sent in on time, it meant that the entire family was not under threat of having to leave the country. One of the most uncomfortable conversations of my professional career was letting my department chair know that unless the situation was resolved within the next 48 hours, I would not be returning to work. I would have to return to Canada and wait there until an entirely new visa application could be processed. I did not end up having to leave—but it took a hefty expedition fee and the goodwill and good humor of a number of staff at both UNCC and the visa office to get things resolved in such a short time span.

Border crossings we have found even more stressful than visa issues. As a family we have learned to turn up at the airport much earlier than recommended. Before we got our green cards, border agents seemed to enjoy extended conversations with us about our work and our thoughts about why we had been issued work visas in the United States. "How does your degree qualify you to do what you do?" one asked. Easy answer for me (all three degrees in geography); less easy for my husband, it seemed. Apparently, a degree in creative writing and film bears little connection to a profession in publishing. I always felt sorry for the people who were in the line behind us. While this experience of delay is a common transnational one among visa holders, less usual is having a family member whose name and passport combination appear on the "no-fly" list. The first time my husband was detained was a year after 9/11. We were returning to the United States having spent Christmas holidays with my parents in Toronto. After asking the usual suite of questions, the agent asked my husband to gather his family and follow him through an unmarked, obscured set of glass doors at the far side of the immigration hall. We did as we were told. Once through the doors we were immediately separated. There followed no explanation of why we were there or where they had taken my husband. I was left with my two children (the youngest an infant) alone in a sterile waiting room. After what seemed like an eternity, another officer came into the room to tell me that as soon as they completed their process with Interpol and confirmed that my husband was not the person they were looking for, we would be free to go. It would be a little while yet, he said. A less frightening version of that scenario continued for three or four years afterwards and, after an unexplained hiatus, seems to have begun again this year.

I share these personal stories of border crossings and visa challenges not only because they are an essential part of my transnational experience, but also because I think they shape my positionality as an immigrant scholar. My personal immigrant story is textured by a world of waiting, of writing large checks to lawyers and governments, of relying on the bureaucracy of employers to do what needs to be done, of deadlines and due dates, of shifting visas and rules, different citizenships, passports, and visa statuses for different members of my family, occasional visits to the homeland security office for photo documentation, fingerprints and eye scans, of delays and discomfort at the border, and realizing that you are a part of a much larger system under which you have very little control. While complicated, expensive, and sometimes frustrating, my experience is benign compared to the experiences of the many immigrants to America and Canada with whom I have worked as part of my professional research and outreach efforts. But it has provided me with a shared point of reference and a bridge into sometimes difficult conversations. I know a little something of their relief when they or a loved one are finally granted a green card or extension of status. We share the common experience of living out our daily lives at a physical and emotional distance from some of our most important family members. I can empathize when they talk about missing home or are not able to return there because of visa restrictions or status issues. With that all said, there is a magnitude of difference between their experiences and mine. I did not have to flee my country in the dark of night for fear of persecution because of my political beliefs. I did not have to walk for days across the desert borderlands. I did not as a child have to ride atop a train through some of the most dangerous provinces of Guatemala or Mexico to get to El Norte. I do not have to send remittances back home to help build a house or a freshwater well for the family I have left behind. I know where my parents, my children, and all of my siblings are. The courage and bravery of so many of the people with whom I have worked over the years humbles me beyond words, and it always reminds me powerfully, and at times painfully, of my privilege.

A LONGING FOR HOME

Even now, I get homesick.

The sound of a Canadian accent, the copy of Maclean's magazine on my brother-in-law's coffee table, the soaring voice of KD Lang, my daily reading of the online version of the *Globe* and *Mail*, the taste of freshly baked apple pie with sharp cheddar cheese. Small things, big impact.

I have been at UNC Charlotte for sixteen years. When I accepted my first tenure-track position, I expected it to be just that—the first of others to come. I always assumed that I would, at some point, make the move back home. But the enduring and unanswered question for me, as it is for so many immigrants, remains "Where is home?" My parents still live in the same house in which I was raised, and the view from their front porch remains as breathtaking as ever. But, they are only there at certain times of the year—often they are with me and my family or with my sister and her family who, in an odd twist of fate, also settled in North Carolina. My parents, on occasion, will lament being left behind. Both their children emigrated from Canada to the United States. I respond by saying that I think our relationship is stronger, richer for being so many thousands of miles away. We cherish and make the most of the time we spend together, and my children are fortunate beyond measure to have grown up with one set of their grandparents living with them for months at a time over the summer and early winter months of every year. With regularity, my parents will also say that if the cost of health care in the United States was not a barrier, they would join us permanently. Because this barrier exists, a growing concern of mine is how as a family we will navigate the challenging years to come when the health issues of advanced age inevitably arise and my parents are not as independent, active, and mobile as they are now. What will happen when they can no longer travel to us, or no longer drive a car or get to appointments and activities with ease? When someone needs to be with them for support, caregiving, friendship? My transnationalism is our transnationalism. My parents are affected by their adult children's move away from their country of birth, and they too live a life that must negotiate the border in complicated and emotionally fraught ways.

So why I am not *home*? In Charlotte, I have built a professional and personal life in the United States that I believe could not be replicated in Canada. Both my husband and I were raised in global cities, and we gravitate personally and professionally to the urban. The Canadian cities in which I would most want to live are, at the moment, financially out of reach. And regardless of where they might be located, job openings for tenured full professors of geography are rare, to say the least. Many years ago, I had the opportunity to return to Canada, but when all was said and done, an uncertain job market for my husband, a lower salary for me, no guarantee of tenure, the exorbitant costs of housing and everyday living made the move financially and professionally unwise. It remains the hardest decision I have ever had to make.

But it was the right one. At UNC Charlotte I have had the great privilege to work with outstanding students and tremendously generous colleagues both within and outside of the academy. Together we have explored the rapidly changing immigrant and ethno-cultural landscape of Charlotte, North Carolina, and the broader New South. Committed to a brand of scholarship that links theory with practice and the building of interventions that flow from what we have learned, I have worked in partnership with local advocates, medical practitioners, researchers, and members of the documented and undocumented immigrant community to use the tools of geography to build programs that ensure primary care access for newly arrived Latino immigrants and help them settle into, and become an integral part of, their new communities. With a local museum, I have been involved in the development and evaluation of exhibits and dialogue programs that help both receiving and arriving communities understand and adjust to rapid cultural change and work together toward a city of greater integration and inclusivity.

FUZZY TRANSNATIONALISM, FUZZY IDENTITY

In her work on American migrants in Toronto, Vancouver, and Halifax, Susan Hardwick explores the mixed feelings of citizenship and belonging that her interviewees shared (Hardwick 2010). Through their stories, she illustrates that transnationalism is temporally and spatially contingent, and spatially and temporally variable. Transnationalism's practice and experience shifts over time and across space. When I reflect upon the transnational nature of my life, I think of it in these fluid, fuzzy terms. While I am most certainly Canadian, and identify strongly as such, home for me is now in two countries—Canada and the United States. The ingredients that make each place home, however, differ. The United States is the country of my midlife. It is where, together with my husband, we have raised our daughters, purchased property, built a house, crafted careers, nurtured friendships, engaged civically, and constructed community. It is where we have invested ourselves. In 2013 my British-born husband, Canadian-born daughter, and I became American citizens.

Canada is the country of my childhood and transition into adulthood—it is the foundation upon which the structure of my life has been built. My ideals and values clearly reflect my Canadian upbringing, and my optimistic nature is a by-product of both how and where I was raised. These traits also underpin how we are raising our daughters—one Canadian, the other American. Irrespective of their place of birth, they know that an important part of their heritage and culture is Canadian (and, to make things fuzzier still, given my husband's birthplace, also British). A very joyful part of our transnational practice is the celebration of two Thanksgivings, one celebrated the second Monday of October and the other the last Thursday of November. I have continued the tradition of opening our table to a group beyond immediate family. Last year at Canadian Thanksgiving, we had twenty-three people crowded around the six-seat dining room table which was once my grandmother's. Conversation at this annual event inevitably turns to what it means to be Canadian. I have often fumbled for an answer.

As many Canadians will attest, the answer most readily within reach is: to be Canadian is to *not* be American. But certainly there is more to this than a flip response that provides for lively dinner conversation. Canadians, in my experience, are on the whole more liberal, more tolerant of difference, more embracing of diversity, more inclined to care for fellow citizens, kinder and more humane in their government, as evidenced by their universal health care system. Such characteristics do not entirely vanish when one crosses the border, and they certainly didn't disappear when I took my U.S. citizenship oath. Nor are these traits unique to Canada or her citizens. I live and work among Americans who, in practice and belief, are just as, if not more so "Canadian" than I.

As my family deepens roots where we are now, at work, at schools, at clubs and organizations, as dual (and in the case of my husband and eldest daughter, triple) citizens, our citizenship is formalized and acknowledged, yet somehow we still remain a step removed from where we live now and from where we once came. I do not think that my children feel this way, but my husband and I often talk about only partially belonging in either place. At times, we feel like we are global citizens in a world that, in practical terms, simply does not allow for such things, belonging in multiple places and nowhere at the same time. Our citizenship, our belonging, our identity is fuzzy indeed.

I became a geographer all those years ago because I wanted to make sense of the complexities, and unexpected similarities, of the different kinds of places and communities I was encountering as a young and naïve Canadian living, exploring, and studying in the United States. While I think I have made some progress on that front, it may be some time yet before I can fully understand the complexity of my own transnational identity.

REFERENCES

Hardwick, Susan. 2010. Fuzzy Transnationals? American Settlement, Identity, and Belonging in Canada. *American Review of Canadian Studies* 40 (1): 86–103

Hardwick, Susan. 2014. "Canadian and American Cross-Border Migration, Settlement and Belonging." In *Transnational Borders, Transnational Lives*. 7–22. Eds. Remy Trembley and Susan M. Hardwick. Québec: Presses de Université du Québec.

Hardwick, Susan and Heather Smith. 2012. "Crossing the 49th Parallel: US born migrants in Canada and Canadians in the United States." In *Immigrants in American and Canadian Cities*. 288–311, Eds. J. C. Texiera, W. Li and A. Kobayashi. Toronto: Oxford University Press.

Ley, David. 1988. Styles of the Times: Liberal and Neo-conservative Landscapes in Inner Vancouver, 1968–1986. *Journal of Historical Geography* 13 (1): 40–56.

Lucas, Susan. 2006. Canadian Transnationalism: Canadian Living and Working in the United States. *The Geography Teacher* 3 (2): 38–40.

Mills, C.A. 1988. Life on the Upslope: The Postmodern Landscape of Gentrification. *Environment and Planning D: Society and Space* 6 (2): 169 –190.

Mills, C.A. 1991. Fairview Slopes: Gentrification in the Inner City. *Canadian Geographer* 35 (3): 306–310.

Ornstein, M. 2006. *Ethno-racial groups in Toronto, 1971–2001: A Demographic and Socio-Economic Profile*. Toronto: Institute for Social Research, York University.

Rossiter, D. 2014. "Crossing the Line: Reflections of a Transnational Lifeworld." In *Transnational Borders, Transnational Lives*. 41–52. Eds. Remy Trembley and Susan M. Hardwick. Québec: Presses de Université du Québec.

Swanson, K. 2014. "Where is home? An Autoethnography of Academic Migration." In *Transnational Borders, Transnational Lives*. 27–39. Eds. Remy Trembley and Susan M. Hardwick. Québec: Presses de Université du Québec.

Index

About the Contributors

Heike Alberts is a professor of geography at the University of Wisconsin Oshkosh. She was born and raised in Berlin, Germany, and came to the United States as an international student in 1998. She completed her PhD at the University of Minnesota and began teaching in Oshkosh in 2003. Her research interests focus on international migration and urban development. She loves to travel and spends several months each year in Germany.

Stavros T. Constantinou is an associate professor of geography at Ohio State University, Mansfield Campus. His interests include the study of migration and ethnicity from the quantitative point of view. His published research includes two major studies on emigration from Greece to the United States and emigration from Cyprus to the United Kingdom (with Nicholas D. Diamantides) as well as a study of Greek American ethnic identity (with Milton E. Harvey). In addition to survey data, he has used census and church records to examine the changing spatial dimensions of Greek populations in the United States.

Ibipo Johnston-Anumonwo is a professor in the geography department at State University of New York, College at Cortland. She has a bachelor's degree in education from the University of Ibadan in Nigeria, a master's degree from the Harvard Graduate School of Education, and a PhD in geography from Clark University, Worcester, Massachusetts. A 2010 recipient of the SUNY Chancellor's Award for Excellence in Scholarship, and the AAG 2016 Distinguished Teaching Honors recipient, she serves on the editorial boards of *African Geographical Review* and *WAGADU: Journal of Transnational Women's and Gender Studies;* and she previously served on the editorial board of *Gender, Place and Culture,* and on the Advanced Placement Human Geography Test Development Committee. As a member of the executive board of the New York African Studies Association, she served as president and co-edited two books on African studies. Her latest book, *Global Perspectives on Gender and Space* (Routledge), is co-edited with Ann Oberhauser.

Weronika A. Kusek is a human geographer focusing on Polish migration and diaspora. Weronika grew up in Poznan, Poland. She came to the United States to attend the University of Toledo. She received her PhD from Kent State University. She is currently an assistant professor at the Department of Earth, Environmental, and Geographical Sciences at Northern Michigan University. Weronika is a recipient of the AAG Hoffman Award (2013), and she chairs the AAG Cultural Geography Specialty Groups.

Wei Li received her geography BS and MS degrees in Beijing, China; and her PhD in geography at the University of Southern California. She is a professor at the Asian Pacific American Studies/School of Social Transformation and the School of Geographical Sciences and Urban Planning at Arizona State University. Her foci of research are highly skilled international migration, integration, and transnational connections, focusing on the Pacific Rim. She is the author or co-editor of five scholarly books and two journal theme issues, and has 119 other academic or educational publications. Funding sources of her work include the National Science Foundation and Canadian government. She was a member (2003–2012), Vice Chair (2004–2010), and Chair (2010–2012) of the U.S. Census Bureau's Race and Ethnic Advisory Committees (REAC) on the Asian Population; a member of the International Steering Committee for the International Metropolis Project; and the North American Director for the International Society of Studying Chinese Overseas (2010–2016). She also served at the inaugural class of the National Asia Research Associates with the National Bureau of Asian Research and Woodrow Wilson International Center for Scholars, was a Fulbright Scholar to Canada (2006–2007) and India (2016–2017), and was a recipient of the Rockefeller Foundation's Bellagio Writing Fellowship (2014).

Alan P. Marcus is an associate professor at the Department of Geography and Environmental Planning, Towson University, Towson, Maryland. He is a broadly trained geographer, with regional specialization in Brazil. Areas of research include: transnationalism, immigration, race, and Brazilian immigration. His work has been published in several anthologies, as well as in professional journals such as: *Annals of the Association of American Geographers*, *Geographical Review*, *Journal of Cultural Geography*, *Journal of Latin American Geography*, and *Journal of Immigrant and Refugee Studies*.

Heather A. Smith is professor and director of Geography Graduate Programs at the University of North Carolina at Charlotte. She also serves as director of the College of Liberal Arts and Sciences' interdisciplinary Urban Studies Minor and is past president of the AAG Ethnic Geography Specialty Group. Dr. Smith's research and teaching explore the intersections between transitioning societies and restructuring cities. Most recently, she has explored immigrant settlement and receptivity in non-traditional gateways; the dynamics of Hispanic "hyper-growth" across the U.S. South; and the complex relationships between Latino settlement, immigration status, and community and individual health. She is co-editor of *Latinos in the New South: Transformations of Place* (Ashgate 2006) and of *Charlotte, NC: The Global Evolution of a New South City* (Georgia 2010). Dr. Smith's long-standing commitment to participatory research and engaged scholarship was recently recognized with UNC Charlotte's 2014 Provost's Faculty Award for Community Engagement.

Shaolu Yu is Mellon Post-Doctoral Fellow in Urban Studies at Rhodes College. She holds a BA in resources, environment, urban planning and management (Qufu, China), an MA in urban geography (Beijing, China), and a PhD in human geography (University of Connecticut). Trained as an urban geographer in an interdisciplinary background, she has developed a comparative and global perspective and a mixed method approach in her research on cities. Her dissertation titled "Transnationalism, Mobility and Identity: the Making of Place in Flushing, Queens, NewYork City" examines the paradox of transnational mobilities at the cross-national scale and immobilities at the local scale of the Chinese immigrants in an urban environment. The project also investigates how the paradox is embedded within individual Chinese immigrants' everyday routine of life, and how it impacts their identities and sense of place. Her current research interests include: transnationalism, race and ethnicity, urban segregation and ethnic enclaves, transportation geography and mobility, and Asian urbanism. She is also interested in applying cartography and GIS in visualizing and analyzing spatial data.

www.ingramcontent.com/pod-product-compliance
Lightning Source LLC
Chambersburg PA
CBHW051251200926
51612CB00001B/2
9781498509480